Dictionary of
Applied Energy Conservation

Dictionary of
Applied Energy Conservation

David Kut

BSc(Eng), CEng, FIMechE, FCIBS, FInstE, MConsE

Kogan Page, London
Nichols Publishing Company, New York

First published 1982 by Kogan Page Ltd
120 Pentonville Road, London N1 9JN

British Library Cataloguing in Publication Data

Kut, David
  Dictionary of applied energy conservation.
  1. Energy conservation – Dictionaries
  I. Title
  333.79′16′0321 HD9502

  ISBN 0-85038-577-6

Library of Congress Cataloging in Publication Data

Kut, David
  A dictionary of applied energy conservation.

  1. Energy conservation – Dictionaries. I. Title.
TJ163.28.K87 621.042 82-3565
ISBN 0-89397-131-6   AACR2

Printed in Great Britain by
The Anchor Press Limited
and bound by William Brendon
both of Tiptree, Essex

Preface

The escalating cost of energy is drawing an ever increasing number of people into the planning and execution of energy conservation measures and programmes and confronts them with the specialist terminology of the conservationist. The object of this illustrated dictionary is to list the generality of terms employed in energy conservation practice and to explain, with the aid of appropriate illustrations, the basic definitions and underlying techniques.

I trust that this publication will prove of valuable assistance to all concerned with saving energy, be they architects, industrialists, managers or lay public.

David Kut, *August 1982*

Acknowledgements

The drafting and final presentation of this illustrated dictionary has been an extended and demanding task; I acknowledge the valuable assistance given by my secretary, Mrs Sheila Merlin, and by our senior engineer, Mr G Glover, who has so ably dealt with the illustrations.

Picture Acknowledgements

The author and publishers would like to express their thanks to The Architectural Press Ltd for permission to reproduce various illustrations from their books. The sources of the illustrations are as follows: *Applied Solar Energy;* figure numbers 1, 2, 7, 10, 11, 47, 71, 103 to 106, 118, 119. *Waste Recycling for Energy Conservation;* figure numbers 3, 8, 9, 13, 18 to 24, 31, 37, 39, 40, 48 to 52, 56, 68, 69, 73 to 75, 77, 89 to 93, 96, 100, 102, 107, 114. *District Heating and Cooling for Energy Conservation;* figure numbers 17, 25 to 30, 32, 36, 45, 46, 57 to 59, 61, 62, 64, 65, 70, 72, 76, 78, 79, 85, 94, 98, 101, 111 to 113, 115.

Bibliography

In compiling this illustrated dictionary I have
necessarily drawn on many sources, but I wish to
acknowledge particularly the information I have
derived from the following books:

*The British Gas Directory of Energy: Energy Saving
Equipment 1980-81.* Cambridge, UK: Cambridge
Information and Research Services, 1980.

Diamant, RME, and Kut, David: *District Heating and
Cooling for Energy Conservation.* London:
Architectural Press, 1981.

Kut, David: *Heating and Hot Water Services in
Buildings.* Oxford: Pergamon Press, 1968, 1976.

Kut, David: *Warm Air Heating.* Oxford: Pergamon
Press, 1970.

Kut, David, and Hare, Gerard: *Applied Solar Energy.*
London: Architectural Press, 1979.

Kut, David: *Applied Waste Recycling for Energy
Conservation.* London: Architectural Press, 1981.

absolute air filter See *air filter – absolute.*

absolute humidity See *humidity – absolute.*

absolute temperature The temperature at which there is no heat energy left in a body (– 273 °C). Expressed in degrees Kelvin. To convert conventional temperature to absolute temperature, add 273.6 to Centigrade temperature (456.7 to Fahrenheit temperature).

absorptance The ratio of the radiant energy absorbed by a plane surface to the radiant energy incident upon that surface. (See figure 1.)

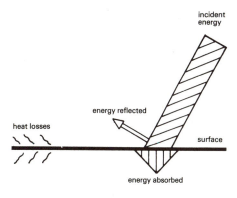

Figure 1 Absorptance

absorption The process by which radiation is converted within a material into excitation energy. Most of the radiant energy falling (or incident) upon a matt black surface is absorbed.

absorption chiller Equipment utilising the principle of absorption refrigeration. See *absorption refrigeration.*

absorption refrigeration Substitutes a heat source (gas firing, steam or medium pressure hot water) for the compressor in the conventional chiller. The heat vaporises the absorbent, usually lithium bromide; it is then taken through a cycle of operation which includes concentrating and condensing the absorbent, evaporation of the liquid and spraying of same into an absorber, from which the dilute solution is returned to the concentrator. Alternative to compressor operated vapour refrigeration system. Reduces operating costs where cheaper heat sources are available (possibly surplus steam during the summer air-conditioning season).

Relatively costly and space consuming with associated cooling tower plant. (See figure 2.) *Application:* general use; in particular for energy recovery where there is available surplus heat.

absorptive attenuator Attenuator that incorporates glass fibre and mineral wool materials, and that is effective over a wide range of frequencies.

accumulator A vessel in which heat is accumulated during a period of slack demand for heat and discharged at times of high demand.
  Steam: Useful in circumstances where a relatively high proportion of the total steam demand is required or met at a pressure which is substantially below the steam main from the boiler and where steam is at times in surplus and available for storage.
  Application: waste heat system associated with incinerators where the incineration process is continuous and the steam demand fluctuating; wherever there are advantages in maintaining constant steam production or back-pressure steam feed in a situation of varying steam demand.
  It has been found practicable in many cases to adapt redundant Lancashire boilers as accumulators.
  Hot water: Large, efficiently insulated storage vessel in which the hot water is heated during periods of surplus steam availability for use of hot water at times of high demand for hot water.

acid condensation The deposition of acid out of acid-containing gas or vapour when cooled below the dew point of the acid. Relates particularly to chimney flue gases; acid condensation can cause major damage to chimney and associated installation and therefore must be avoided.

acid dewpoint Refers particularly to chimney operation in which the flue gases have an acid content (eg sulphur). Cooled below the acid dew point, the acid will condense out of the flue gas.

actimatic grease interceptor See *grease interceptor – actimatic.*

activated carbon filter Activated carbon granules located in position by a glass fabric on either side of the air filter panel. The filter assembly is then placed into the air stream which is to be filtered, offering adequate area to air flow.

Activated carbon has the ability to absorb vapours or gases in adequate capacity for purposes of commercial air filtration. This ability derives from the nature of its surfaces, which are made up of myriads of sub-microscopic pores.

Activated carbon filters are widely employed for the removal by absorption of body odours, tobacco smells and fumes, cooking smells, etc. Only a relatively brief period of contact is required between the activated carbon and the air being filtered.

The carbon material employed in air filters is selected for its properties of fine porosity, absorption and durability to resist abrasion and crushing during manufacture and handling.

When handling fairly dirty air, as is commonly the case in cities and in industrial areas, a fabric type or other suitable pre-filter must be placed into the air stream upstream of the activated carbon filter to reduce the dust burden on that filter, leaving the filter to deal mainly with vapours and gases.

The state of the activated carbon filter medium is usually monitored by a test panel fitted to each filter section. The panels are removed after a specified period of service and are subjected to a laboratory test to assess the likely life of the main filter medium in the circumstances of a particular installation.

Activated carbon filters are particularly useful in applications where a high proportion of ventilation air must be recirculated, where cooking smells can cause offence, etc.

activated silica A highly effective coagulant aid which is prepared from sodium silicate 'activated' by various chemicals.

active system Relates to solar energy collection. Refers to a system which utilises *external* energy (commonly electricity or possibly gas) in the collection and/or exploitation of the solar energy; for example, hot water system which incorporates a circulating pump; warm air collector and distributor which rely on fan(s) and electric motor(s); heat pumps employing compressor, condenser fan and automatic controls.

adiabatic process A process during which no exchange or transfer of heat occurs between a system and its surroundings. It can only occur if (1) no temperature difference exists or (2) the system is thermally insulated from its surroundings. These qualifications indicate that a true adiabatic process cannot be achieved in practical engineering, as there is no known perfect heat insulator. By controlling the temperature of the immediate surroundings, so that it always equals (or almost equals) that of the system, a very close approach to an adiabatic process can be achieved.
Application: compression of a gas in a refrigeration compressor – adiabatic compression may be assumed in computing the theoretical power requirement of the process.

adsorption A process caused by the mass

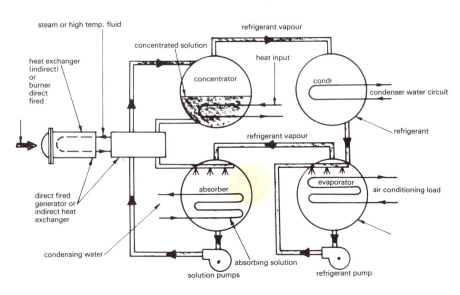

Figure 2 Absorption refrigeration diagrammatic arrangement of process

migration of the vapour molecules from the air to the adsorbent because of a higher vapour pressure in the air than exists within the pores of the adsorbent substance.

Some adsorbents are selective relative to the vapours they will attract into themselves. Eg silica gel, when exposed to a mixture of vapours, will invariably adsorb the water vapour from the mixture; charcoal will ordinarily adsorb odourbearing vapours. Thus silica gel is well suited to dehumidification applications; charcoal to the purification of air and odour control.

a/e ratio The ratio of absorption coefficient for solar radiation to the emission coefficient at operating temperature; it is a measure of the selectivity of absorber surfaces.

aeration The practice of exposing small droplets of water to air to encourage the absorption of oxygen from the atmosphere or the release of other gases in order to effect beneficial changes in the water.

aerobic corrosion See *corrosion – aerobic*.

aerodynamic noise generation In general, caused by air (at high velocity) flowing around or against surfaces (such as duct splitters, elbows without guide vanes, dampers blades, diffusers, grilles, etc) and resulting in turbulence. Noise can be re-radiated from an attenuated section to break out in an unattenuated section of the duct system.
It is of particular concern and importance in modern high-velocity air systems, with air flowing above 10m/s and at pressures of up to 20mb.
Can be avoided or rectified by careful design of ducted air systems and by selective application of sound attenuation.

aerogenerator A machine for converting wind energy and electric power. Growing number of applications in Europe and USA of projected outputs up to 4 MW per unit. UK Central Electricity Generating Board plans wind farms with individual outputs of up to 1MW fed into the electricity grid. Modern aerogenerators of large output are mounted atop towers up to 100m high and may have blades of up to 40m long.

after-cooler Associated with air-compressor systems. Fitted between air compressor and air receiver. Cools the compressed air to ambient temperature after compression; this permits majority of the water vapour in the air input to condense and be removed before it enters with the air into the receiver. Usual cooling medium

is water. Common design basis: 5.5°C (10°F) temperature rise of the cooling water; air temperature cooled within 11°C (20°F) of the water inlet temperature. Oil and moisture must be drained off periodically. Presence of these pollutants can be a major factor in the failure and breakdown of control equipment of the compressed air system.

airborne thermal infra-red heat loss survey
Airborne thermal infra-red line-scan equipment for locating energy losses from external surfaces of buildings, storage vessels and pipelines due to structural or material defects. Surveys are carried out during the winter heating period.
Application: energy conservation schemes for manufacturing plants, chemical works, hospitals, public buildings, etc.

air-circulating ovens Use an air-recirculation system as opposed to the traditional approach of expelling the air after one pass through the working chamber. A heat exchanger can be sited in the burner exhaust duct if required.
Application: storing, curing and heat treatment ovens or any other process oven requirement.
Energy-saving potential: the air-recirculation system reduces heat input by up to 60 per cent over a dead-loss system. Additional benefit is obtained with the use of the heat exchanger.

air classifier Employed in waste separation and recycling to direct the different separated materials into their respective discharge streams. Basically, comprises a blower, a rotating drum fitted with an internal helix and a large engagement chamber. In use, a current of air is directed along the drum, and the sized feed is allowed to fall through the air stream, so that any paper and plastic in the mixture are carried along the drum and into the disengagement chamber. The dense objects fall through the air stream and are conveyed to the opposite end of the drum by the internal helix. (See figure 3.)

air conditioner Term applied to an assembly of air-conditioning equipment, usually within one package.
Packaged: Compact air-conditioning unit in which all the essential components are assembled into one unit.
Split type: Package air conditioner in which the evaporator is in its own cabinet; the compressor and condenser (the relatively noisy and vibratory) equipment is usually located externally to the air-conditioned space, but connected to the evaporator cabinet by insulated pipes. Such arrangement provides a

quieter air-conditioning environment and facilitates layout planning. Capacity range up to about 5kw.

Feasible to have the compressor packages with either the evaporator or the condenser; the latter is more usual.

air-conditioner – window type: Packaged air-conditioner assembly fitted into window of air-conditioned space, so that condenser faces to the outside. Permits simple installation. *Disadvantage:* relatively high noise level in air-conditioned space (depending on model) and unsightly external appearance.

air conditioning (comprehensive) Provides control of temperature (heating and cooling), relative humidity, air condition (filtration) and air movement within the treated environment.

air conditioning – fan-coil system See *Fan-coil air-conditioning system.*

air conditioning – induction system See *Induction air-conditioning system.*

air conditioning – two duct system See *Two-duct air-conditioning system.*

air conditioning – variable volume system See *Variable volume air-conditioning system.*

air conditioning Versatemp system See *Versatemp air-conditioning system.*

air controller – rhomboidal See *Rhomboidal air controller.*

air curtain A cold or warm air curtain with adjustable blast direction control. Single side, top blow or double side blow units are available. Wind velocities up to 15 mph, depending on conditions, can be resisted. Warm air curtains available from which warm air from door heater is discharged directly into the work area

as a prime source of heating. A door-actuated switch directs the air blast across the door only when the door is open. *Application:* to prevent escape of warm air from heated space.

air diffuser Duct system terminal fitting which discharges air into a space. Consists of a number of metal cones with the air outlets between the cones.

Each configuration provides a characteristic air-flow pattern. The designer must select same to suit the particular application.

Diffusers may be circular or rectangular (square).

air filter Assembly of dust-, gas- or dirt-absorbing surfaces fitted into a ducted system to clean the air passing through it. The filter sheet metal housing is usually provided with flanges for attachment to ductwork. There are many types of filter and filter media to suit the various filtration requirements encountered in research and industry.

air filter – absolute Fabric dry-type filter unit suitable for filtration at ultra-high efficiency down to 0.01 micron particle size. Will remove dust, bacteria, virus concentrations, etc. Employs as filtration medium a special form of glass-based paper containing a proportion of finely corded, long asbestos fibre or similar material.

Application: hospital, clean room, research, etc situations.

air filter – bag Comprises a series of filter bags within the filter assembly. Can be made to provide a high-filtration efficiency by suitable specification of filter medium. Can incorporate automatic dust-shaking device. Can handle large dust load.

air filter – ceramic Used in the removal of acid mists and in compressed air filtration. Suitable

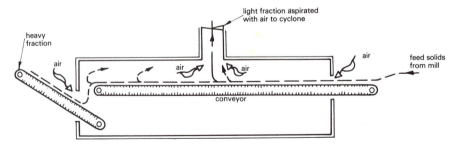

Figure 3 Principle of the enadimsa horizontal air classifier

for use on high-pressure and high-temperature applications (500°C (932°F)).
Can be made suitable to withstand thermal shock.

air filter – electrostatic Has three major components: ionizer, dust collector and electronic power pack. Ionizer comprises a series of earthed tubes; between adjoining tubes is stretched a fine metal wire charged to about 13,000 V direct current (d.c.). When the ionizer wire is charged positively, the amount of ozone present in the filtered air is not increased beyond that found in the atmosphere at sea level on a normal sunny day.
A strong electrostatic field is set up across the space between the ionizing wire and the earthed tube surface, and, at the correct voltage, a corona discharge takes place from the wire, causing ionization of the air molecules, which are thereby greatly accelerated. When a dust particle carried by the air stream passes through this space, it is met by a stream of ions travelling outwards from the ionizing wire. Collision with this stream ionizes the dust particles and causes these to become impelled in the direction of the collector cell, which consists of a number of flat, parallel, vertical plates; one set is earthed and the second set charged to approximately 6000 V. The plates are arranged alternately so that the air and the dust flow along narrow passages with an earthed plate on one side and a charged plate on the other. The electrostatic field between the plates in the collector cell draws the dust particles to the earthed plate, to which they adhere until removed.
Cleaning of the electrostatic filter is carried out by washing down the cells with warm water; it is often found convenient to install sparge pipes for this purpose. The permissible length of time between cleaning operations depends on the location of the plant, the proportion of air recirculation and the season of the year (much more frequent cleaning than normal would be required during periods of a heavy London fog).
Sumps with adequate drain connections and water seals must be provided, the object of the seals being to prevent ingress of unfiltered air via the drains. Man-access to the filter is through air-tight doors with safety locks, interlocked with the power pack to prevent opening of the access doors unless the electric supply to the filter unit has been switched off. (See figure 4.)

air filter – fabric-type Uses a fabric filtration medium (as opposed to liquid, metal, etc). Suitable for filtration down to 2 micron particle size. Material must be suited to the application and may have to be flame and fire resistant in certain instances.

air filter – gauge Indicates filter resistance. Can incorporate warning lights.

air filter – grease filtration type Used in conjunction with cooking and similar equipment which generates suspended grease. Is usually located at the outlet(s) of the exhaust canopy where this connects with the extract ducting. Filter cells are usually provided with means of easy handling and withdrawal for cleaning in suitable detergent solution and re-use.

air filter – pre-filter A coarse type of filter used ahead of main filter. Essential to protect

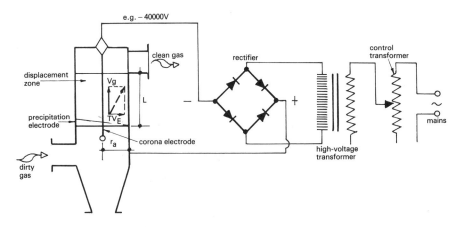

Figure 4 Electrostic air filter-diagrammatic arrangement

absolute, carbon and electrostatic main filters; otherwise employed where dust burden is great.

air filter – renewable-type The filter medium may be cleaned by compressed air or vacuum, or it may be a viscous liquid which is recycled through a suitable liquid cleaner.

air filter – roll-type Filtering material is wrapped around a drum which automatically advances to maintain an acceptably clean surface of filtration medium. An end-of-roll warning alarm or light indication must be provided to alert the maintenance staff to the need for roll replacement. Tends to provide relatively coarse filtering.

air filter – throwaway-type The filter medium is expendable. When saturated with the filtered material or when its resistance has become excessive, the material is removed, thrown away and replaced with new. The filter casing must have facilities for monitoring filter performance and must be provided with means of easy withdrawal. (See Figure 5.)

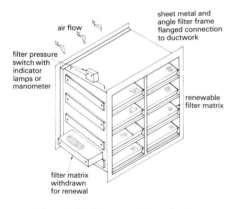

Figure 5 Air filter assembly – showing withdrawal facility

airfilter – vee-type Filter cells arranged in V formation inside the assembly to offer the greatest surface area to the air flow.

air flow – in ducts Rate of air flow in a duct system depends on:
 resistance to air flow offered by air ducts and associated duct fittings;
 density of the air being conveyed;
 absolute viscosity of the air;
 internal dimensions of ducts and fittings;
 shape and air flow characteristics of the system components;

roughness of the duct wall.
Guidance on design air-flow systems is provided in *IHVE Guide,* section C4 (CIBS – Flow of Fluids in Pipes and Ducts, 1977).

air flow measurement Field measurements of air flow are required to establish that the completed system meets the design intent and to achieve a balance of air flow around the various sections of a ducted air system.

air governor – automatic Used in conjunction with air compressors. Provides for the continuous running of the compressor, but allows the plant to run under conditions of light load when a predetermined pressure has been reached in the air receiver. The driving unit (eg electric motor) will then be running on light load with a corresponding reduction in power consumption. When the demand for air increases, the pressure in the air receiver drops and the automatic air governor allows the compressor to resume operation. A hand-lifting device is often incorporated and enables any air under pressure from the air receiver to lift the governor valve and so pass into the unloading cylinder of the compressor (this ensures that the air compressor can be started up against frictional load only, even if air receiver and pipe lines are full of air under pressure).

air-handling luminaire Dual-purpose light fitting which embodies within one fitting the light source (tungsten or fluorescent) and a facility for supplying or extracting ventilation/air-conditioning air through the fitting. Extraction is more common and can be associated with energy-recovery system in which the heat given off by the light source is withdrawn into the extraction plant before it can raise the space temperature; this raises the temperature of the extracted air and suits it that much better for heat recovery via heat pump, recovery heat exchange coil or heat wheel.

air lift Method of raising water from a low to a higher level by the injection of compressed air.

air lock 1. Accumulation of air trapped in a pipe or vessel sufficient to reduce or stop the flow of the working fluid.
2. Space sealed by two doors to reduce the air change rate when moving from one set of air conditions to another.
3. Space arranged to prevent loss of conditioned air when entering or leaving.

air mass The length of the path through the earth's atmosphere traversed by direct solar radiation, expressed as a multiple of the path length with the sun at zenith.

air – primary See *Primary air.*

air receiver Associated with air compressor equipment. Damps out the pulsations of the air delivered by the compressor, stores the air to provide a supply 'cushion', cools the air and removes the oil and moisture which is contained in the input air (usually in association with ancilliary equipment).
Each air receiver must be permanently marked with:
   maker's identification number;
   date of test;
   specification number;
   hydraulic test pressure;
   maximum permitted working pressure.
Air receivers must be installed, maintained and periodically inspected under the provisions of the Factory Act. B.S. Nos. 429 and 487 relate to air receivers.

air requirement for combustion See *Combustion air requirement.*

air – secondary See *Secondary air.*

air-supply diffuser May be circular, square, rectangular or of slot-type. Permits good control over air-flow pattern. Variety of dampering methods available, eg by screwdriver inserted from front, quadrant control at rear,

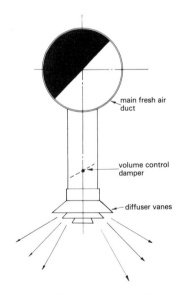

main fresh air
duct

volume control
damper

diffuser vanes

Figure 6 Circular air supply diffuser

local adjustment at each slot diffuser. Engineered to suit particular circumstances and requires careful selection. (See figure 6.)
Application: for all types of air-conditioning and ventilation systems. Very large diffusers available for use in tall and large spaces.

air-supply nozzle Nozzle-shaped air-supply fittings suitable for applications which require a long air throw from the discharge. Tend to be noisy.
Application: air supply into large and tall areas, such as exhibition halls, factories.

air vent A device for removing air from a liquid system, either manually or automatically.

air vents for steam The air vents open to discharge air or an air/steam mixture and close against live steam. This is done automatically and irrespective of steam pressure.
Application: the main application of the equipment is to continuously remove air; for example, air vents work on steaming ovens, ironer beds, drying cylinders.
Energy-saving potential: eliminating 6 per cent of air in the steam supply to a unit heater can increase efficiency by 30 per cent. Removing air from a jacketed boiling pan can reduce cooking time by half. The automatic action of the vents enables machines to reach operating temperature rapidly, which can save fuel and avoid poor production.

air washer Employed with ventilation/air-conditioning systems. Sprays water into the air system to humidify, dehumidify or cool. May rely on evaporative cooling or operate with chilled water. Incorporates pump(s) and controls.

alarm monitor Related to underground (eg district heating, steam, condensate, etc.) insulated pipes using a small diameter cable (wire) which is incorporated into the insulated pipeline. In the event of a leak or of damage, the resistance of the cable changes and releases a visual or an audible alarm in the plant room or district heating station. Thermographic (infra-red photography) techniques can then be used to locate the exact position of the damage. Some pre-insulated pipe systems incorporate an alarm monitoring system.
Major advantage: early warning of damage; this should permit remedial action before the damage becomes more widespread and costly to repair.

alternative energy Term assumes that the fuels

(oil, gas, coal) are the sources of *conventional* energy. Other sources of energy are termed *alternative*. These include wave and wind power, solar, bio-gas, nuclear, methane, alcohol, hydrogen, geothermal refuse, etc.

altitude The angle which the rays of the sun make with the horizontal plane at a given point. (See figure 7.)

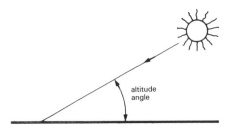

Figure 7 Altitude

alum Aluminium sulphate, a widely used coagulant.

aluminium cladding Term applied to external metal chimney insulation; relates to the application of a skin of aluminium sheet over an air space or over thermal insulation of a vertical steel chimney. Method of application is laid down in British Standards Specification B.S. 4076:1978.

aluminium foil – self-adhesive .002 in. thick aluminium foil self-adhesive coated and protected by a peel-away backing paper for easy application; supplied in easily handled rolls.
Application: wide-width material is used primarily for backing radiators. Narrow-width material is used for sealing joints in foil-faced duct and pipeline insulation.
Energy-saving potential: applied to the wall behind radiators (reflective side out), it will show a substantial decrease in heat loss through a wall. As an insulation joint sealer, it makes insulation fully efficient.

ambient Surrounding (temperature).

ambient noise The existing background noise in an area. Can be sounds from many sources, near and far.

amplitude The peak value of any sinusoidal function.

anaerobic digester for bio-gas generation The material to be digested is stored in a tank for the appropriate period, heated and stirred. The recovered gas is held under a floating or rigid cap as required and is available for use in boilers or gas-engine-driven electric generators. The digested slurry can be used after further treatment for fertilizer or food protein recovery. (See figure 8.)
Application: for use with biodegradable industrial wastes and farm and abattoir slurries.
Energy-saving potential: the energy recovered from the digester can be used as heat, electricity or indirectly as fertiliser and food protein. It also reduces the cost of effluent disposal.
Availability: 16 to 24 weeks.

anaerobic digester – general Animal or other organic wastes treated in the digester under controlled conditions and at temperatures of 35°C (95°F) are biologically broken down to produce a mixture of methane gas and carbon dioxide in the approximate proportions of 70:30. Sufficient methane is obtained to operate the plant and provide surplus power to run other on-site services. (See figure 9.)
Energy-saving potential: the energy available from digestion of various organic wastes and materials varies between 0.3 and $0.8m^3$ of bio-gas per kg volatile solids.

anchor See *Pipe anchor*.

anemometer Hand-held vane (windmill) operated instrument for measuring the velocity of air movement.

angle of repose Natural running angle or natural angle of slope of a material (eg coal) at which a bulk store of the material will settle. Natural angle of repose of major coal types: dry broken coal – 33°; dry slack coal – 30°; dry coke – 35°; coke breeze with 5 per cent moisture – 40°.
Important in the design of coal-storage hoppers and bunkers; a minimum angle of 50° is recommended to establish a smooth flow of the fuel to the handling equipment.

anode – sacrificial See *Sacrificial anode*.

anti-freeze solution Prevents freezing, usually in water, by various additives or by heating trace elements keeping fluid temperature above freezing point – 0°C (32°F).

anti-reflection coating (abbreviated as **AR**) A coating applied to a surface (bloomed) to increase the amount of light penetration. May be applied to the surface of solar cells or to glass/plastic covers of solar collectors.

aquifer – geothermal See *Geothermal aquifer.*

array See *Solar array.*

arrestor A mechanical device which arrests or separates grit and dirt from flue gases. It is usually in the form of a high-efficiency cyclone, working on the principle of a centrifuge.

asbestos Common name for a family of inorganic, fibrous, silicate minerals that possess a chrysotile structure. All grades are fire resistant and have a varying degree of acid resistance coupled with good mechanical strength.

asbestos – blue The crocidolite group of asbestos, mainly mined in Bolivia, South Africa and western Australia. Can be identified by its rich lavender blue colour.
It is held that blue asbestos presents a greater health hazard to workers engaged in its manufacture and to users than the other grades.
Blue asbestos has been used in the past very widely for thermal insulation (largely within asbestos magnesia plastic site-applied insulation). Because of the risk to health, this asbestos is now only rarely used in the UK, but there still

exist many installations which incorporate this material.

asbestos – chrysotile The most abundant of these fibrous silicates; it belongs to the serpentine group of rock-forming minerals and is the most extensively used asbestos for industrial applications. Mainly mined in Canada, the Soviet Union and Zimbabwe.
Grades with relatively long fibres are spun into a yarn, which is used in asbestos cloth weaving for protective clothing, thermal insulation, resin-impregnated pads, wire braiding and ropes.

asbestos – crocidolite See *Asbestos – blue.*

asbestos – health hazard Asbestos is a dusty substance. It has been proved in recent years that this dust can be highly injurious to persons who inhale it in sufficiently large quantities. Asbestos can cause a specific fibrosis of the lungs – now termed asbestosis – largely associated with workers who have been exposed to heavy concentrations of asbestos dust over a long period of time.
A relationship also appears to have been established between exposure to certain types of

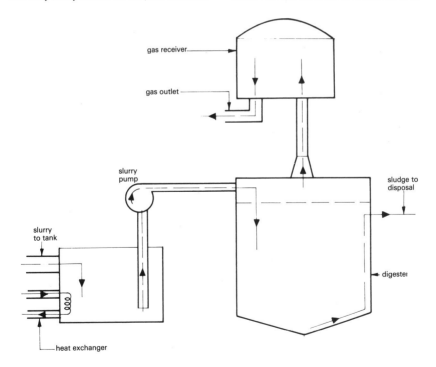

Figure 8 Biogas digestor system

asbestos dust and the occurrence of mesoth-elioma – a form of cancer – after only relatively short periods of exposure. Regulations have therefore been introduced in the UK aimed at greatly minimising this health hazard.

asbestos materials See *Asbestos – varieties of.*

asbestos – regulations Dated 1969, they have been introduced in the UK and supersede the earlier regulations formulated in 1931. They apply to every process which involves asbestos

and to any article which is composed wholly or partly of asbestos, excepting only processes in which asbestos dust cannot arise.

For the purposes of the regulations, asbestos is defined as any of the minerals crocidolite, amosite, chrysotile, fibrous anthophyllite, as well as any mixture containing any of these materials.

The purpose of the regulations is to lay down strict rules for the manufacture, handling and removal of asbestos materials to minimise any health hazard. Indeed, the regulations have re-sulted in the emergence of asbestos-removal

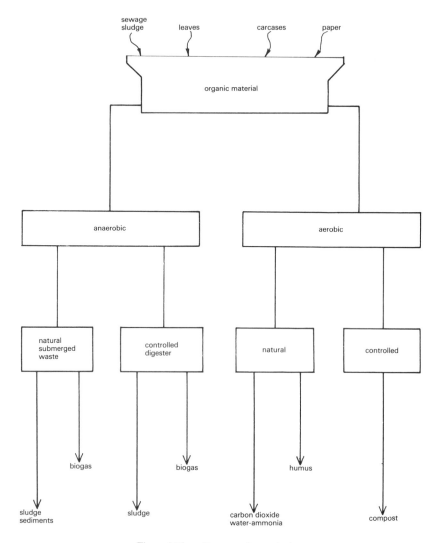

Figure 9 Flow diagram of organic decay

specialists and in a major increase in the cost of asbestos removal from existing installations.

The Health and Safety Executive have published booklet No. 44 under the title *Asbestos: Health Precautions in Industry* (Third Impression, 1977), which provides essential and lucid guidance in this matter.

asbestos use – code of practice Issued in the UK by the Health and Safety Commission; became effective on 1st October 1981. Titled *Approved Code of Practice and Guidance Note – Work with Asbestos Insulation and Asbestos coating.*

Gives practical guidance with respect to the Asbestos Regulations 1969 (S.I. 1969 No. 690) and the Health and Safety at Work Act 1974. Relates particularly to the precautions to be observed in any work involving asbestos-based thermal and acoustic insulation and sprayed coatings in compliance with existing legislation. Endeavours to ensure that exposure to all forms of asbestos dust is reduced to the minimum that is reasonably practicable.

asbestos – varieties of Some 30 or more minerals of fibrous crystalline structure comprise the asbestiform group, but only six have any economic significance; in order of importance, they are chrysotile, crocidolite, amosite, anthophylite, tremolite and actinolite. Chrysotile is a fibrous form of serpentine; the other five comprise the amphibole group.

The name 'asbestos' given to these minerals does not distinguish among its natural types; The two main groups being serpentine and amphibole asbestos. The mineralogical differences are a result of the varieties of matrix in which asbestos occurs. The chemical basis of all types of asbestos is magnesium silicates combined with lime or alkalies in varying amounts.

ash A powdery residue left behind after the combustion process – not to be confused with slag or clinker.

ash content Ash content of a fuel determines partly the calorific value and the required rate of ash removal. The fusion temperature of the ash is important in combustion practice; also knowledge of whether ash solidifies and forms clinker or whether it falls freely.

ashpit Space below solid-fuel and waste-fired furnace grate for accumulation and removal of ash.

aspect ratio (AR) The ratio of the dimensions of two adjacent sides of a rectangular duct.

For optimum design, the aspect ratios of rectangular ducts should be as low as practicable to minimise friction losses and thereby economise in fan input energy.

atmospheric gas boiler Chimney-induced operation (without fan) usually complete with integral draught diverter, copper heat exchanger and burner assembly designed for easy cleaning and maintenance. Typical modular design output in 60, 90, 120 and 150 kW sizes.
Application: commercial and large domestic central heating. Models available with non-ferrous headers for launderette use.

attenuation Noise reduction.

attenuator That which provides 'attenuation'. In acoustics, this may be a partition, a piece of lined duct or even air itself. The word is most commonly used to describe factory-made sections of ductwork incorporating large volumes of sound absorbent material and is more correct than the word 'silencer', which implies total sound reduction.

autarchic Self-sufficiency. When related to a building indicates that this is independent of all main services.

autoclave Pressure vessel used for the sterilisation of foods and pharmaceutical articles. Larger sizes are fed with steam from boiler plant; smaller units have integral gas or electric steam generation.

Process autoclaves usually operate on a pre-set programme of pressure and vacuum operation. They are major consumers of steam.

For optimum fuel efficiency, correct steam trapping and pressure controls must be applied and maintained in good order.

Autoclave may be front-operated or be arranged for front loading and rear discharge to best suit its use.

automatic control – floating control The final control element is moved gradually at a constant rate towards either the open or closed position, depending on whether the controlled variable is above or below the neutral zone. The valve (or other controlled element) can assume any position between its two extremes as long as the controlled variable remains within the values which correspond to the neutral zone of the controller. When the controlled variable is outside the neutral zone of the controller, the final control element moves towards the corrective position until the valve of the controlled variable is brought back into the neutral zone of the controller or until the

final control element reaches its extreme position.
Major advantage: gradual load changes can be counteracted by a gradual shifting of the valve position.

automatic controls Form an integral part of each environmental control system. May be self-acting, electrically operated or pneumatically (air) operated.
It is essential that each control application exactly suits the circumstances of use of the controlled system and be as uncomplicated as possible to achieve the control object and function.

automatic control – two-position Controlled equipment (such as a valve) moves quickly between two fixed positions (high or low; open or closed).

automatic differential pressure condensate controller Quickly expels condensate air and other insulators directly they become present. Where these must be lifted to the drain, the unit combines the advantages of blow-through and syphon drainage without the self-imposed disadvantages of either system.
Application: to improve the efficiency of heated rollers and roller dryers in the paper and corrugating industry.

automatic doors Electro-mechanical and electro-hydraulic sliding and swing doors with automatic operation.
Application: for entrance and internal doors in airports, hotels, supermarkets, hospitals, offices, warehouses, old people's and disabled persons' establishments.
Energy-saving potential: typical installations give a 50 per cent increase in traffic flow either allowing for a smaller opening to be used or resulting in the door being in the closed position more often.

automatic ignition Method of initiating the combustion of a fuel by an electric spark, thereby avoiding the use of pilot flame or manual ignition by match, taper, etc. Must incorporate a comprehensive set of fail-safe safety equipment which will cut off the incoming fuel if satisfactory ignition (flame) has not been established within a pre-set time (a few seconds). See also *Flame-failure device.*

automatic lighting control Detects photo-electrically when the natural daylight has fallen to a predetermined level and switches on the electric lighting, which is switched off when the natural daylight increases above the level of the artificial lighting.
Application: interior lighting installations in factories, offices, public buildings, etc.
Energy-saving potential: energy savings of up to 30 per cent are possible.

automatic stoker Device for automatically and continuously feeding solid fuel into a boiler at optimum efficiency. Most common examples of automatic stokers are the chain grate stoker and the underfeed stoker.

automatic voltage stabiliser and load shedding transformer Auto-transformer with on-load tapping switch which automatically controls the voltage to a set level and has an overload facility to reduce the voltage as dictated by a maximum-demand controller. Stabilises voltage supply and automatically sheds a percentage of load if the M.D. (Maximum Demand) setting will be exceeded.

auto-transformer – starter See *Electric motor starter – auto-transformer.*

auxiliary burner Stand-by plant, eg oil or gas burner or any other plant item which is not fully employed in the normal operation of the system.

auxiliary energy The use of an alternative energy source, such as solar, lessens the amount of primary energy (from gas, electricity, etc.) otherwise required. The primary energy provision required is referred to as the auxiliary or back-up need.

axial fan – bifurcated Arranged with electric motor outside air stream handled by fan.
Application: systems handling hot or contaminated air which would damage the electric motor if in contact with same, eg kitchen equipment extract, fume cupboards. Usually can handle hot air at temperatures up to 343°C (650°F).

axial fan – multi-stage contra-rotating Two or more impellers in one casing, rotating in opposite directions. Permits high fan pressure development with pressures up to 50mb.

axial fan – single impeller Fitted with one impeller.

axial flow fan Air enters axially and is discharged axially. Arranged for direct insertion into ducting; comprises impeller with aerofoil section with directly coupled electric motor rotating inside a circular diameter casing which is fitted with flanges for attachment to

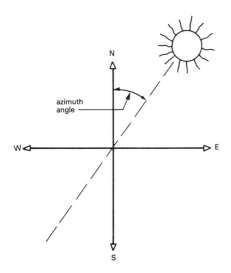

Figure 10 Azimuth

adjoining ductwork.
Advantages: high efficiency, offers choice of blade angle to provide different outputs from same fan size, direct insertion into ducting, overall size for a given duty substantially less than for a centrifugal fan.

azimuth The angle between the horizontal component of the rays of the sun and the true south. (See figure 10.)
Azimuth angle is usually measured in degrees east (morning) and degrees west (afternoon) of south.

23

balancing valve system Employed in the commissioning of heating and chilled water systems through mass-flow measurement. Method uses two valves in the circuit to be balanced: that in the flow pipe for isolating *and* as fixed orifice across which the pressure drop is used as a metering signal; that in the return pipe is of the double-regulating pattern and is used for isolating *and* regulating purposes, the double regulating feature avoiding the necessity for subsequent rebalancing.

Pressure differential signals are indicated at the orifice valve, which, being a fixed orifice, has only one flow curve, and its accuracy is influenced by a minimum of manufacturing tolerances. The flow characteristics of the double regulating valve are of no consequence, other than that they must provide fine regulation and effective isolation.

A given pressure differential signal at the orifice valve results in a positively known mass flow of water. The resistance offered by the circuit being balanced does not matter and may vary from the designer's estimate. When the double-regulating valve has been adjusted to provide a specific signal at the orifice valve, the required (balanced) flow of water will be passing through the circuit.

Each orifice valve should be (and usually is) supplied with pair of PSA/DOE approved test plugs, complete with captive blank caps fitted to each orifice. In commissioning use, probe units are connected to these plugs via plastic tubing to a portable pressure differential test set.

baling – automatic Preparation of material for compact storage using a baling machine; typically bales of previously compacted waste with plastic sealed moisture-proof bags. Bags are then stored without risk of the material crumbling or absorbing moisture.
Application: for dry materials; dry storage facility.

ball valve Controls the automatic water replenishment or make-up in a water tank or similar container. A ball valve usually comprises a metal or plastic float linked by a lever-arm to a slide valve mechanism of a (mains) water valve. (See figure 11.)

ball valve – delayed action See *Delayed-action ball valve.*

barrel (oil) Commercial measure of a quantity of oil – 159 litres. Term used mainly in the oil exploration industry.

baseboard heating USA equivalent of skirting heater (UK). This type of heater is located close to the floor and takes the place of architectural skirting. Comprises a finned heater element (pipe) within a sheet-metal casing of low height.
May be of radiant or convective type. The

B

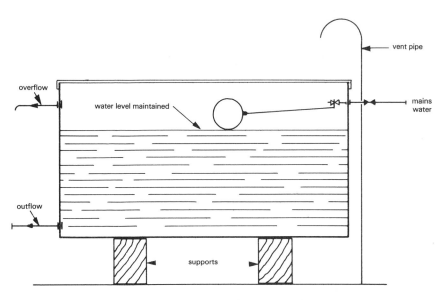

Figure 11 Tank with ball (float) valve

former offers a radiant panel to the room; the latter has a gap at the bottom and outlets at the top to permit a measure of convection. Convection-type heaters tend to cause staining and should be preferably located under glazing only. Thermal output from radiant baseboard is low.

Baseboard heating is best suited to under-glazing installation to counteract draught from the glass surface.

base-exchange water treatment A method of softening hard waters in which a natural or synthetic material introduces sodium-based salts to replace those of calcium and magnesium, softening the water which is passed through it. Has to be regenerated at frequent intervals by washing with a brine solution.

base-line process Flow sheet of process.

baseload That loading of a network or plant (electricity, heat, etc) which is constant. Loadings above the baseload are subject to fluctuations.

bellows expansion compensator Employs the action of stainless steel corrugated bellows to permit unstrained expansion and compression of pipelines. Must be part of a carefully designed compensatory arrangement which will include anchors and pipe guides.

bellows joint – fully articulated Bellows used in sets of two with intermediate pipe of mild steel. Restrained lengthwise with ball-ended tie bars, allowing offset movements in all directions at right angles to axis. (See figure 12.) May be constructed of 10/20 convolutions of single-wall stainless steel for applications in chilled water, low-pressure and medium-pressure hot water, steam and condensate, and oil installations.

May be constructed of 5/4 convolutions of heavy-wall hot formed bellows for applications in medium-pressure and high-pressure hot water, steam, oil and compressed-air installations.

bellows joint – gimbal A variation of the hinged bellows joint. It is hinged at its quadrant points and is therefore able to absorb expansion in *any* plane. Applied to their best advantage, gimbal joints are used in pairs to absorb the pipe movement in planes other than those directed by the pipeline itself.

bellows joint – hinged Bellows joint fitted with hinged frame for applications involving higher

pressures and large angular movement. The two halves of the hinge are each attached to one end of the joint and are pinned at the centre to permit free angular movement restricted to one plane about the centre of the joint.

Hinged bellows joints are more usually employed in sets of two or three. The extent of the lateral movement which such a double-hinged bellows arrangement can accommodate is related to the amount of angular movement the bellows can handle and to the length of the pipe between them.

The permissible lateral movement may be increased by extending the length of pipe between the two hinge pins and also – within the design criteria – the number of convolutions of each bellows. (See figure 12.)

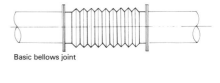

Basic bellows joint

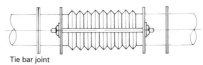

Tie bar joint

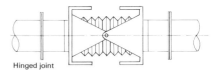

Hinged joint

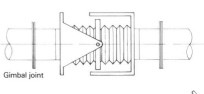

Gimbal joint

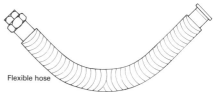

Flexible hose

Figure 12 Pipe-line expansion joint types

bellows joint – neoprene rubber compensator
Constructed of single outward-facing reinforced neoprene convolution with mild steel backing flanges (to suit the pressure condition of the system).
Applied at pumps and similar equipment to smooth out vibration and to permit some movement. Suitable for use with certain chemicals, effluents, brine and oils, as well as with chilled and cold water, low-pressure hot water and hot water service installations.

bellows joint – tie bar Ties the bellows by an arrangement of bars which, for normal single-plane movement, takes the form of hinged bars with the hinges at the centre line of the bellows. For duties at low pressure and for any movement at any plane at right angles to the pipeline, tiebars are fitted with spherical washers under the endnuts to permit angular movement.
In the event of failure of the anchor point, the bellows will be compressed (rather than extended) with resultant increased safeguard to the pipe installation.

belt-driven centrifugal circulating pump Comprises separate pump and electric motor in one assembly, the connection between pump and motor being by a set of pulleys and drive belts. Pump is provided with stuffing box and gland. The larger pumps have a constant dribble of water from the glands which must be piped to a suitable outlet.
A belt guard must be provided. The arrangement can be designed for silent operation. The belt drive permits adjustment of pump performance (within the characteristic of the particular pump) by changing pulley sizes; this may involve also a change of electric motor.

Bernoulli's theorem For any mass of liquid in which there is a continuous connection among all the particles, the *total head* of each particle is the same. This statement is most important in the solution of problems relating to the flow of liquids. The venturi flow meter is based on this theorem.

bifurcated fan See *Axial fan – bifurcated*.

bin discharger Usually fitted within the coned section of a cone-bottomed storage silo to rotate in an angular or a radial fashion in either a clockwise or an anticlockwise direction to agitate the stored waste in the bottom of the silo and thereby encourage the free flow of the material into the associated rotary valve. A manual over-ride is usually incorporated with the bin discharger controls to permit the operation of the discharger independently of the rotary valve, in the event of bridging conditions when fuel starvation arises.

bioenergy process Transforms effluents from food and other factories into a fuel gas. The plants are fully enclosed and do not produce bacterial aerosols, odours or noise.
Application: the dairy industry, fermentation (yeast, beer, cider, pharmaceuticals), distilling and paper and pulp industries.
Energy-saving potential: the process transforms over 90 per cent of organic pollutants into gas with each tonne of organics yielding the equivalent of half a tonne of fuel oil. It is claimed that large distilleries can be made independent of fossil fuels and that cheese factories with a whey disposal problem can be made energy independent of both electricity and heating fuels.
Availability: survey and laboratory services available to ensure reliable design data. Full-scale plants take nine months to build and put into operation.

bio-gas Generated in anaerobic methane fermenters, which convert the organic acids and alcohol into methane and carbon dioxide, their relative proportions depending on the detention time and on the process temperature.
Application: conversion of solid organic wastes; eg manure and other farming wastes are good feedstocks for the generation of bio-gas, which can be used for heating purposes or for driving of an electricity generator.

black body 1. Ideal substance which absorbs all the radiation falling upon it and emits none. 2. A body which emits the maximum possible radiation, ie its emissivity is 1.0.

black heat Relates to air heating applications (mainly electrical) in which the temperature of the heater element(s) is controlled to remain below the red (visible) spectrum, mainly for reasons of safety.
Application: duct-located electric warm air heaters.

blackness test Method of assessing dust-collecting efficiency of air filters. A sample of a test gas is drawn through a filter paper at a definite volumetric rate of flow for a definite time; the resulting stain on the filter paper is then treated for purposes of comparison as the standard. Various samples are taken at the exit side of the filter at the same volumetric rate of flow, but over different time periods. The resultant stains at the exit are then instrument compared with the standard stain. For example, if a certain stain is obtained in 5 minutes on the inlet side of the filter, and a time of 100 minutes is required to achieve a similar stain at the exit of the filter, the efficiency by blackness test is 95 per cent.

bleeder valve Relates to air compressors. Enables the air in the delivery pipeline from the compressor to be exhausted to atmosphere when the pressure switch contacts open and cause the compressor driving motor to stop. May incorporate adjustable pressure differential. May be centrifugally, magnetically, mechanically or oil operated.

blocking diode Allows an electric current to flow through it in only one direction. Incorporated with solar electric systems to prevent batteries discharging through the solar cells when their output is too low to charge the system.

block storage heater Stores off-peak electricity in the form of heat contained in special heat-storage bricks which are heated during the off-peak charging period and discharge heat subsequently. Bulky and heavy.
Application: space heating of offices where occupation is limited to conventional office hours.

block tariff In this a fixed number of electric units is charged and consumed at a comparatively high price before the unit price is reduced to a lower charge.

blowdown See *Boiler blowdown.*

blowdown controller See *Boiler blowdown controller.*

blowdown pit Provides for safe discharge of steam boiler hot water blowdown into the drains. Arranged to store the hot blowdown water to permit its cooling before discharge. Must be fitted with substantial cover to safeguard boiler plant personnel from steam emission.

blown fibre loft and wall insulation Loose glassfibre granule cavity wall and roof space insulation.
Application: domestic and industrial.
Energy-saving potential: savings can be achieved of up to 80 per cent of heat loss through uninsulated roofs, up to 70 per cent

through cavity walls and up to 90 per cent through single-skin walls.

BOD (biochemical oxygen demand) An indicator of the degree of pollution in rivers and watercourses. The oxygen content of water decreases as the amount of sewage present increases. The BOD is the oxygen absorbed at 20°C (68°F) over five days. Most rivers selected for public supplies have BODs of 4mg/1 litre or less.

boiler Term given to a group of equipment in which water is heated by the application of an external heat source to provide a supply of hot water or steam. A specialised type of boiler plant operates with a manufactured heat transfer fluid, usually to provide a liquid at high temperature and moderate pressure.

boiler – air leakage Uncontrolled leakage of air into the boiler via gaps in the brick-setting, defective door and inspection cover seals, unsound joints in the flue system, etc must be obviated by careful workmanship during assembly of the plant and by subsequent good monitoring and housekeeping.
The ingress of unwanted air into the boiler unit depresses the thermal efficiency, as this air must be (wastefully) heated; ingress of air into the flue system will reduce the buoyancy of the draught and, in extreme conditions, can overload the chimney flue gas carrying capacity.

boiler – baffles Cast-iron baffle plates designed for insertion into the flue-way at the junction of sections to lengthen the path of the gases for certain sizes of boilers when these are oil- or gas-fired. Boiler manufacturers generally supply baffles for their makes of boilers and specify their required number and location for the various boilers in the range.
Related to sectional boilers.

boiler blowdown Process of discharging a quantity of boiler water from a steam boiler to maintain the dissolved solids in same within specific limits to avoid carry-over, foaming or priming. Frequency and quantity of blowdown must be carefully monitored and controlled to ensure minimum safe blowdown quantities. Excessive blowdown will waste heat and fuel – possibly in meaningful quantity. Blowdown process must be carried out with care and to specific procedure to safeguard the operator.

boiler blowdown controller Automatically continuously monitors the boiler water density

and adjusts the rate of blowdown to maintain the density within pre-set limits to minimise the amount of the water blown down and consequently the heat loss therein.
Some evidence that the replacement of manual blowdown methods by automatic ones in the average boiler installation reduces the blow-down quantity by about 20 per cent.

boiler – cast iron See *Cast-iron boiler*.

boiler – combination See *Combination boiler*.

boiler – combustion efficiency The energy conversion by the boiler (or other heat generator) firing equipment, expressed as a percentage:

$$100 \times \frac{\text{Heat output (to combustion chamber)}}{\text{Heat input (calorific value or fuel)}}$$

boiler compound Sealing paste used to seal gaps between contact faces of the boiler combustion system.

boiler – corner-tube See *Corner-tube boiler*.

boiler – Cornish See *Cornish boiler*.

boiler damper air sealing Prevents unnecessary heat loss while a boiler is not being fired. Fitted to the burner combustion fan inlet, the air sealing damper prevents cooling air being drawn through the boiler by natural draught during stand- by or shut-down, helps maintain boiler pressure and increases boiler availability. Construction is of heavy galvanised steel complete with servo motor and interlock. Butterfly damper and synthetic rubber seal design is claimed to achieve 99.95 per cent sealing.

boiler – economic See *Economic boiler*.

boiler economiser A stack-mounted economiser designed to recover energy from the exhaust gases for pre-heating of boiler feed water or process usage.
Energy-saving potential: savings of 3 per cent to 9 per cent on boiler operating costs.

boiler – electrode See *Electrode boiler*.

boiler – hot-water Term for group of boilers which supply heated water for space heating, process or domestic hot-water purposes. Boilers may operate at low pressure (temperature below boiling point of water) or at high pressure (temperature above boiling point of water).

boiler – hot-water/high temperature Has primary object of providing water at a higher temperature than can be achieved by an open system or by minor pressurisation. Temperature tends to be limited to about 180°C (356°F). Such high temperature can only be achieved by operation at high pressure. See also *High-pressure/medium-pressure hot-water system*.
Boilers must be manufactured and tested to the appropriate working pressure; this involves higher standards and greater costs than for low-pressure hot-water boilers.

boiler – hot-water/low-temperature Commonly used for residential and commercial space heating and domestic hot-water provision. In most cases, boiler operates under pressure (head) provided by an open expansion cistern at the topmost part of the system connected off the boiler. Boiler system may also be sealed by use of pressure vessel. See also *Sealed system*.

boiler jacket Thermal insulation within a metal protective envelope. Usually factory applied to packaged boilers and supplied in sections for site erection to sectional boilers.

boiler – La Mont See *La Mont boiler*.

boiler – Lancashire See *Lancashire boiler*.

boiler – liquid phase See *Liquid phase boiler*.

boiler – modular See *Modular boiler*.

boiler mountings Relates to set of pressure-, temperature-, combustion- and draught-indicating and safety equipment which is mounted directly upon the boiler.

boiler – packaged See *Packaged boiler*.

boiler plant – roof-top See *Roof-top boiler plant*.

boiler rating Heat or steam output from the boiler specified by the manufacturer for a particular boiler type and model; usually given in terms of kw/hr (Btu/hr) or kg/hr (lbs/hr). In the case of steam boiler, rating is tied to a particular output pressure and feed water temperature.

boiler return water boost pump Recirculates proportion of the flow water off the boiler to the return boiler entry to maintain a temperature differential across the boiler of about 22°C (40°F) in order to avoid acid corrosion at the return entry point. Particularly important when burning fuels with high sulphur content.

boiler – sectional See *Sectional boiler.*

boiler sequence control Method by which the sequence in which steam or hot-water boilers and boilers within a group of boilers are switched on or off is related strictly to the requirements of the heat or process loads/demands.
Energy-saving potential: by matching boiler on-line output to the heat demand, the boiler plant can be operated more economically than with random switching – possible saving to 10 per cent.

boiler sequence controller Electrical mechanism for effecting sequence control of multiple-boiler installations.
See also *Boiler sequence control.*

boiler setting Relates to the refractory brickwork within a boiler or furnace.

boiler smoke – tubes Tubes which carry the hot combustion and flue gases through the boiler water content and which effect heat transfer.

boiler storage capacity Relates directly to volume of water within the boiler, which acts as heat storage. Boilers with large water content have a higher storage capacity than boilers with low water content. The former have more inherent stability but greater weight; the latter are more difficult to adjust relative to the associated water circuits but have less weight and occupy less space.

boiler – thermal efficiency The energy conversion by the overall boiler unit (or other heat generator), including losses, e.g. via chimney, radiation, infiltration:

$$100 \times \frac{\text{Heat output from boiler to connected system}}{\text{Heat input (calorific value of fuel)}}$$

boiler tube cleaning equipment Electric, air-driven or manual tube cleaning machines; tools and brushes for all types of tubular boilers and process plant.
Application: tube cleaning: removing scale and waste combustion products from the inside of boiler tubes, etc.
Energy-saving potential: equipment usually

pays for itself by the first or second boiler clean.

boiler – wall hung See *Wall-hung boiler.*

boiler wastage Loss of thickness of boiler components due to the effect of corrosion; mainly due to presence of damp soot on and between the boiler surfaces and failure or difficulty related to routine cleaning of boiler combustion surfaces.

boiler – waste-heat See *Waste-heat boiler.*

boiler – water-tube See *Water-tube boiler.*

bonding Applies to pipes and other metal work connected together by a conducting metal bond to ensure common electric potential.

bonding and bonds Term used (in the context of this book) with refractories. The *stretcher* bond is mainly used in smaller furnace construction. The *header* bond is used for furnaces which operate at high temperature. The *English* bond comprises alternate courses of headers and stretchers. The *Dutch* bond is similar to the English bond, but ensures that alternate stretcher courses are not coincident. (See figure 13.)

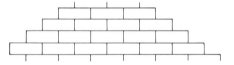

1. Stretcher bond

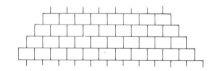

2. Header bond

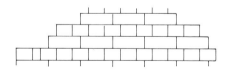

3. English bond

Figure 13 Standard brick bonding methods

borehole – geothermal See *Geothermal borehole*.

bottoming See *Solid state*.

boundary film Layer of stagnant fluid (liquid or gas) at the boundary of a pipe, heat transfer surface, etc. Acts as insulator and offers considerable resistance to heat flow. Insulating effect can be reduced by agitation of the fluid.

bower barffing Process for rust-proofing cast iron or mild steel. The metal is raised to red heat and is treated with live steam.

Boyle's Law The absolute pressure of a given mass of any gas varies inversely as the volume, provided that the temperature remains constant.

braiding Plaited cable protective covering of fibrous or metallic composition.

breakout – noise The loss of sound energy by default from a system designed to insulate or attenuate an excess acoustic energy.

break tank Located intermediately between the mains water supply and the equipment or installation provided with water. Such tanks are required in connection with open-space heating systems, commercial size sealed systems, hose reel installation, submerged inlet bidets, etc.
The water supply *from* the break tank may be pump-boosted or free flowing.

brine Solution of inorganic salt in water used as secondary heat transfer fluid in refrigeration systems (usually sodium chloride or calcium chloride salts are used). Sodium chloride may come into contact (if necessary) with food-stuffs; calcium chloride (which has an un-pleasant taste) may not. Suitable for lower temperatures than chilled water (to about 20°F (-2°C)). Tends to be corrosive.

briquettes storage For housing wood-waste briquettes pending use; must be dry and safe-guard the stored material from crumbling and reverting to its original state.

briquetting Process employs a briquetting machine, which compresses (surplus) wood-waste chips and sawdust into briquettes of convenient size for storage pending demand. Briquetting press operates at high pressure and is noisy. Requires careful location.
Application: timber-processing industries,

where supply and demand for the waste heat are not in balance with each other.

brown-coal See *Coal – brown*.

brush draught excluder Strips of brush-type fibre (eg polypropylene) bonded to a non-corrosive flexible carrier which can be cut to length at random. May include a (polypropylene) fin run integrally through the brush material to improve draught effectiveness and to act as a water barrier. Excluder is cut to required length and is fitted at bottom and sides of external doors to exclude draught.
Application: doors to warehouses, factories, hospitals, loading bays, hotels, etc.

brush strip draught excluder A brush strip made by looping nylon filaments over a core wire. The assembly is then locked into a metal channel and the resultant brush strip housed in an aluminium PVC carrier. Available in a variety of depths.
Application: for sealing all types of doors and wood-framed windows, including sliding sash windows.

Btu Imperial unit of heat (flow) and abbreviation for British Thermal Unit. That quantity of heat which is required to raise the temperature of 1 lb of water through 1 degree Fahrenheit.

Btu meter Equipment designed to integrate and record Btu (or other measuring units) measuring hot-water flow and flow return temperature.
Application: designed for use within industrial boiler plant.

bucket elevator Enclosed inclined or vertical elevator which conveys material by scoops or buckets. Used in conveying of coal in boiler plants.

building energy management systems
Computer-based system for the supervision and automatic operation of buildings and systems. Programs available for peak electrical demand limitation, optimum start/stop, enthalpy optimisation, efficiency monitoring, planned maintenance, actual energy consumption and cost, direct digital control of plant, access control, fire detection, tour patrol, flexi-time, lighting, etc.
Application: large buildings or complexes, chain stores, etc.

bulk density – general Weight density of (waste) fuel expressed in kg/m³. Can vary widely for same fuel; eg wood waste may vary between 80kg/m^3 and double this density, depending on whether it is hard wood or soft wood, whether it is in the form of wood dust, off-cuts or chips, and whether it is dry or wet.

bulk density – waste Density of different wastes stored in bulk; this varies greatly, particularly if the waste is wet. For example, the density of wood waste may vary from about 80kg/m^3 to over 160kg/m^3, depending upon whether it is hard wood or soft wood, whether it is in the form of wood dust, chips or off-cuts, and whether it is moist, wet or dry. The variation of bulk density in other forms of waste may be just as great. The variation of waste fuel volume within a combustion chamber has a significant effect on the combustion air requirements.

burner – auxiliary See *Auxiliary burner.*

busbar A solid uninsulated conductor to which several circuits are connected. Usually protected inside a busbar chamber.

butterfly damper Regulating device inserted into ductwork to control gas air fiow. Comprises single-leaf control from a damper quadrant. Offers crude method of control of leakage.
Application: small ventilation plants.

butterfly valve Regulating device fitted into pipework to control fluid flow. Comprises single-blade control. Offers crude control. Susceptible to leakage. If used for isolation purposes in a circulation with thermo-syphon effect, it must be fitted within the flow pipe to best effect.
Application: residential heating systems; certain process uses.

bypass factor (BF) Relates to cooler batteries in air-conditioning systems. Defined as that proportion of the air passing through the coil that does not *touch* the cooling surface and hence is not cooled by passage through the cooling battery.

calorie That quantity of heat which is required to raise the temperature of 1 g of water through 1 degree Centigrade.

calorific value The quantity of heat which is released during the combustion per unit weight or volume. The intrinsic value of any waste as a fuel, therefore, is directly related to its calorific value. Calorific values of different fuels depend largely on the amounts of carbon and hydrogen they contain, the heat being derived from the combustion of these elements to carbon dioxide and water.

calorific value – higher The gross or upper calorific value, which is the quantity of heat liberated per unit mass of fuel burnt when the products of combustion are cooled down to the standard atmospheric temperature.

calorific value – lower The higher calorific value less the latent heat of the steam, which is generated in the combustion of the hydrogen present in the fuel and in the evaporation of the moisture associated with the combustion air.

calorific values of some typical wastes The following wastes have a significant calorific (fuel) value:

Waste	Approx calorific value (KJ/kg)
Gases	
Coke-oven	43,500
Blast furnace	2,650
Carbon monoxide	1,340
Refinery	50,700
Liquids	
Industrial sludge	9,000
Black liquor	10,230
Sulfite liquor	9,770
Dirty solvents	30,000
Spent lubricants	27,500
Paints and resins	18,500
Oily waste and residue	41,870
Solids	
Bagasse	11,500
Bark	11,000
General wood wastes	13,000
Sawdust and shavings	14,000
Coffee grounds	13,500
Nut hulls	17,900
Rice hulls	13,500
Corn cobs	18,900
Paper and cardboard	17,000
Refuse-derived fuel (RDF)	9,000
Chicken litter	12,000
Mushroom compost	8,000

C

calorifier bundle Assembly of tubes onto a tube plate to provide immersion heating to a liquid (most commonly water) storage vessel. May be suitable for steam, or liquid heat transfer medium. Arranged with flow and return or steam and condensate connections for connecting up to primary heating medium pipes. The storage vessel should be fitted with suitable access/inspection manhole for viewing condition of the bundle.
Heat-transfer surface must be specified to suit the required heat output with reference to available primary heating medium.

calorifier Heat exchanger which comprises two independent heat circuits: a primary one which introduces the heat into the equipment from an external heat source ¬ a secondary one which accepts its heating from the primary circuit. Thus, there are two sets of circuit connections to a calorifier: the primary which services the primary heater and the secondary via which the heated medium (water or other liquid) circulates between the calorifier and the points of heat use.

calorifier – chest Applies to steam-to-water calorifiers. That part of the equipment to which the heater battery is connected inside the calorifier and which incorporates the pipe connections.

calorifier cradles Purpose-made supports for neat installation of horizontally mounted calorifiers, commonly cast in one piece.

calorifier – non-storage type Does *not* incorporate meaningful secondary storage capacity, so that the secondary hot water flows rapidly across the calorifier to the points of use.
Application: space or process heating.

calorifier – specification The designer must specify to the calorifier manufacturer the available primary medium (pressure of steam or temperature of the hot water); the secondary water temperature; the required heat exchange per hour; the static pressure to which the calorifier will be subjected; the various ancilliary connections (for safety valve, vent and cold feed); and the required temperature control.

calorifier – steam-to-water The primary heating medium is steam; this condenses within a heater battery or coil and transmits heat to the secondary hot water circuit. (See figures 14 and 15.)

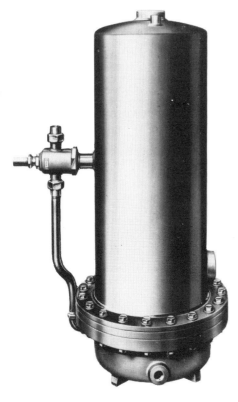

Figure 14 Vertical non storage steam to water calorifier
Note: steam chest and thermostatic control valve in steam supply

calorifier – storage type Incorporates a meaningful capacity for the storage of the secondary hot water. (See figure 16.)
Application: domestic hot water systems.

calorifier – water-to-water The primary medium is hot water at a suitable temperature (commonly high pressure hot water) which heats the secondary domestic or process hot water. The latter must always be at a lower temperature than the primary hot water.

canned rotor circulating pump Centrifugal hot-water circulating pump of lightweight glandless compact construction. Silent operation. Complete unit is made of a single or three-phase electric motor and a pump; designed for high output whilst offering relatively low resistance to the flow of water. Entire rotating element, including the electric motor, is separated completely from the stator and windings by a shell of non-magnetic metal. The bearings are made of stainless steel, and they are lubricated by the circulation of the water; they do not require oiling or greasing. The bearings are not located in the main circuit, but are lubricated by water drawn off from the main circuit along the shaft of the unit. The absence of a stuffing box ensures water tightness.

The most recent such pumps have a graduated integral adjustable output control knob. Such pumps operate so quietly that the provision of a neon light indicator is useful to check that the pump is in operation.

Pump is suitable for domestic dwellings (its main and popular field of use) and for the smaller commercial and industrial installations.

In the event of pump failure, it is usual practice (certainly for the domestic size units) not to effect a site repair, but to exchange the faulty pump for a new one.

capacitance The particular property of a system of electric conductors and insulators which allows them to store an electric charge when there is a potential difference between the conductors. Measured in Farads (F).

capacitor The use of capacitors for power-factor improvement has certain advantages over other methods. Capacitor assemblies are made for all voltages up to 33,000 volts, and can be connected to circuits of any voltage without the use of special transformers. They have a low temperature rise and negligible losses. They occupy little floor space and do not require special foundations. The individual capacitor element consists of continuous bands of metal foil separated by layers of high-grade insulating paper, the whole being wound into cylindrical form upon an insulating tube. The elements are themselves connected in parallel groups, or in series parallel for the higher voltages.

For individual control, capacitors should be located close to the load. They should be connected across the terminals of the induction motor or other inductive apparatus, under the control of the motor switch. When the motor is switched off, the capacitor will discharge itself through the motor windings. For group control, the capacitors are connected to the group busbars, and are controlled by separate switches, fitted with high-ohmic resistances for discharge purposes.

capillary pipe fitting Joints (light-gauge) copper pipe in which the joint is secured by the flow of solder through capillary action along the annular space between the outside of the pipe and the inside of the pipe fitting.

carbon (C) Major combustible component of fuel; the greater the carbon content, the more valuable the energy potential of the fuel.

carbon dioxide (CO_2) Related to combustion of fuel, it is the predominant product of the combustion process. The percentage of CO_2 in a flue gas, coupled with the flue gas temperature, indicates the efficiency of combustion; eg for fuel oil: 11 per cent CO_2 at 260°C (500°F) yields a combustion efficiency of 80 per cent; the same conditions, but with CO_2 of 8 per cent, yields a combustion efficiency of only 75 per cent.

carbon monoxide (CO) Product of incomplete combustion; its presence indicates a deficiency in the energy process.

carbon monoxide monitor Continuously monitors carbon monoxide levels in industrial boilers, kilns, large furnaces, etc using an on-line, non-sampling infra-red absorption technique.
Application: the monitor enables excess air to be minimised whilst avoiding the onset of carbon monoxide and solids emission. Assists in keeping corrosion and pollution to the minimum.

carry-over Also called entrainment. The conveying of particles of water and the constituent chemical salts – in solution or precipitate form – by steam. The principal forms of carry-over are spray, foam and slugs of water. The amount of carry-over may be excessive under conditions of foaming and priming. Increases moisture content of conveyed steam.

cast-iron boiler Available in complete units for small domestic size applications and in sectional construction for medium-size residential and commercial installations.
Boilers available for hot-water and low-pressure steam generation. The steam boiler is fitted with an m.s. steam drum fitted on top of the individual sections.
Due to the limitation of the cast-iron material, such boilers are generally limited for operation under a maximum pressure (head) of about 35m to 40m.
Cast-iron boilers are more resistant to corrosion than m.s. boilers. Can be adapted for direct hot-water supply by treating the waterways with a glass lining. When used for direct hot-water supply, they have fewer flue-passes (and lower efficiency) and are provided with clean-out mudholes.
Suitable for steam generation in closed system only. See also *Sectional boiler.*

catalyst Substance which is introduced into a chemical process to increase the rate of reaction, without the substance (the catalyst)

Figure 15 Horizontal non-storage calorifier
Note: support cradles

itself being materially altered in the process.

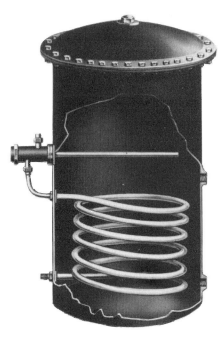

Figure 16 Section through storage type steam to water calorifier for domestic hot water duty Note: heat exchange coil and thermostatic control

catchpit Envelope placed under, or around, leak-prone equipment or container of fluid to contain spillage due to malfunction or over-filling. Catchpits usually provided in association with oil burners and oil fill points (to catch drips) and oil-storage tanks (to guard against spillage of oil into adjacent ground and drainage system); also associated with water-storage tanks in areas where over-filling or bursting could result in severe damage. Catchpit has to be constructed of material appropriate to the fluid involved; catchpits around oil-storage tanks have to be rendered oil-resistant.

cathode Negative electrode. Applies to corrosion protection systems.

cathodic corrosion protection Reduces or elim-inates the corrosion of buried steel pipelines caused by the effects of water penetration to the metal with or without the associated presence of air. Protection is achieved by the provision of a sacrificial anode (a large lump of

steel or other suitable metal) connected to the pipeline to serve as an anode. Corrosion sub-sequently occurs on this sacrificial anode and not on the pipe. Figure 17 illustrates the application of cathodic protection.

cathodic protection Method of guarding metal pipes, apparatus or vessels exposed to corrosion by modifying the corrosive action so that the object of protection becomes the cathode instead of the anode of the system.

caustic embrittlement Relates to a particular cause of boiler component failure. It is a form of cracking in the components which is mainly intercrystalline and occurs with the continued repetition of stress concentrations. When present, it is usually found in areas liable to be under the influence of static stress and concen-trations of considerable magnitude and in contact with concentrated solutions of caustic soda.

cavitation The flashing of hot water into steam which occurs when, due to a pressure reduction, the saturation temperature of the hot water is lowered in a system. High-pressure hot-water systems are prone to cavitation, which may be brought about by the loss of pressure due to pipe friction. A common precaution utilises a fixed-position by-pass valve, which always blends a fixed quantity of return water into the system to maintain a temperature margin between the boiler water temperature and that in the system.

cavity wall insulation Injected into wall structures of cavity construction in the form of a foam to fill the cavity spaces and provide additional thermal insulation to the wall. Suc-cessful injection requires great care and expertise.

C.D.F. Central fuel distribution. Relates spec-ifically to a central system of oil distribution for extensive housing estates in which the oil for the whole estate is kept in one central oil store and is pumped from there through a pipe-line to the individual consumers. Each dwelling has its own boiler and heating installation.

CEGB Abbreviation for Central Electricity Generating Board, which is responsible for the generation of all electric power in the UK and for feeding it into the national electricity grid.

ceiling fan economiser Comprising a thermo-statically controlled ceiling fan in a casing

which blows hot air downwards. Suitable for heights up to 16m.
Energy-saving potential: a 10-12 per cent reduction in heat supply requirements is possible.

ceiling heating by gas radiants Various arrangements of gas-fired radiant panels or direct gas-fired radiant tubes suspended at high level.
Application: industrial.

ceiling heating by radiant strips Arrangement in which a radiant reflector is attached to a single hot-water or steam pipe; thermal insulation is applied above the reflector to direct the heat radiation downwards.
Application: industrial for tall buildings.

ceiling heating – embedded Arrangement of serpentine pipe coils (steel or copper) embedded within the concrete ceiling construction of the building.
The pipe coil forming the panel is placed on the shuttering before the floor is cast; where the ceiling is subsequently plastered, slip tiles are placed between the bends of the serpentine coils as a key for the plaster. Plastering material and its application must conform to specification formulated for this purpose, if subsequent cracking of the ceiling surfaces is to

be avoided.
Where no plaster finish is provided (eg in factories, warehouses, etc), the panels are located on distance pieces above the shuttering, so that a cover of fair-faced concrete is formed below the pipes.
Flow temperature usually limited to 54°C (130°F) to avoid excessive stresses in the building fabric. Pipe spacing and panel location to suit heat losses.
Advantage: low-temperature radiation from the ceiling is an effective and pleasant manner of heating. Cleanliness assured.

ceiling heating – radiant panels Employs special radiant panels mounted at high level above the works or warehouse area. Typical panel has serpentine pipe coil at back and flat metal plate at the radiant output side. May have pipe coil side fitted with thermal insulation where radiation is required in one direction only. May have double-sided flat panels over the pipe coils.
Suitable for use with low- or high-temperature hot water or with steam.
Application: industrial.

ceiling heating – with false ceiling Arrangement of providing ceiling space heating by serpentine pipe coils laid inside a sub-ceiling. System requires thermal insulation above the pipes and contact between the metal pipes and

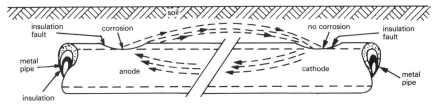

Drawing shows buried pipework without cathodic protection. Note the area of corrosion (anode) when the current leaves it.

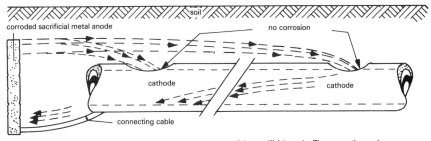

Buried pipework with cathodic protection. Note the corroded state of the sacrificial anode. There are other and more sophisticated forms of cathodic protection.

Figure 17 Corrosion effects with buried pipelines

the special metal ceiling. Ceiling panels may be perforated or plain.
System can operate at general circuit temperature. Radiant ceiling can incorporate modular light fittings. Due to the higher ceiling temperature, the system is unsuitable for rooms less than 3m high.

ceiling panel heating See *Ceiling heating.*

cellular glass insulation A totally inert cellular glass insulation material which is water-vapour, fire, rot and vermin proof, and is dimensionally stable. It is supplied in slab form or fabricated for pipe work.
Application: Flat roofing, cavity wall insulation, pipe cladding and tank bases.

centigrade Temperature scale widely and increasingly used. The unit is 'one degree Centigrade' or 1°C.
See also *Temperature scales.*

centrifugal fan See *Fan – centrifugal.*

centrifugal pump Operates with an impeller which rotates inside the pump casing. Impeller is mounted on pump shaft, which is driven either directly off the drive motor or via a belt drive and pulleys.

ceramic air filter See *Air filter – ceramic.*

ceramic heat wheel A heat recovery system incorporating a ceramic heat wheel for application up to 1,200°C (2,192°F). The system package includes combustion air and exhaust fans, interconnecting ducting, safety controls etc.
Application: for combustion air pre-heating on furnaces in which exhaust gas temperature exceeds 500°C (932°F). Fume incineration. Indirect heating.
Energy-saving potential: the system offers 70 per cent to 75 per cent efficiency in heat recovery across the heat wheel. A fuel saving of 25 per cent is offered by pre-heating air up to 500°C (932°F) from exhaust gasses at 800°C (1,472°F).

chain grate stoker Relates to the automatic firing of a boiler with solid fuel, such as coal, peat and suitable solid waste. Comprises a travelling grate which automatically feeds the fuel into the combustion chamber.
The links of the chain grate are cast in a special grade of cast iron to fine limits to confer a long working life in what is a punishing plant duty. The links are joined by bright steel link rods of

circular section, and the arrangement is so designed that these require no special fastenings or locations to permit speedy replacement of worn links.
Fine air spacing between the links reduces the need for riddling, side cooling of the links and maintains even air entrained riddlings to be deposited on the ash extractor.
The chain grate is driven by a constant-speed, totally enclosed electric motor through an infinitely variable hydraulic gear and reduction gearing to a main worm and worm-wheel driving the stoker front shaft. The speed of the grate is usually variable from 0 to about 7m/hr. If necessary, the grate may be driven in reverse.
As a safety device, a shear-pin is provided to protect the grate mechanism in the event of blockage or overloading. Provision is incorporated for hand operation in an emergency.
The fuel hopper has a full-width shutter; when in the open position, it acts as a fuel cut-off valve. The *depth* of the fuel bed is controlled by a refractory-lined guillotine door. By raising this door, sufficient height is gained for inspection or hand firing.
Chain grate stokers have been developed for the successful firing of refuse-derived fuel (RDF) when pre-mixed with high quality coal in the ratio of not less than 50 per cent by weight.
The chain grate stoker operates in conjunction with appropriate ash handling system. (See figure 18.)

change-over valve system Operates to change the functioning of a plant from one mode to another, eg changing over a two-pipe fan-coil or induction air-conditioning system from cooling to heating; change the heat pump mode from cooling to heating.

Charles's Law A given mass of a gas expands by a constant fraction of its volume at 0°C when its temperature is raised by 1 degree, provided that the pressure remains constant.

check valve Permits the flow of a fluid in one direction only under certain conditions; eg when the pump in a distribution system is switched on or off.

chemical heat pipe See *Heat pipe systems.*

Chemlec cell system Electrolytic recovery system of valuable metals from dilute solution manufactured by Bewt Water Engineers Ltd, UK. (See figure 19.)

chilled water battery Assembly of parallel pipes within a flanged metal casing. Pipes are connected across two headers (chilled water flow and return). The air is cooled by passage through the battery. Temperature control is commonly by thermostat and motorised valve.

chiller – absorption See *Absorption chiller.*

chimney All the essential chimney sections, chimney fittings and accessories necessary for the conveyance of flue gases from an appliance or a flue pipe to the external atmosphere.

chimney – double-skin See *Double-skin chimney.*

chimney connector An accessory which connects an appliance or flue pipe to the chimney. It may be manufactured as part of the support assembly.

chimney draught Difference in pressure between chimney base and terminal; this motivates the evacuation of the flue gases through the chimney system.

chimney draught – balanced Two fans are employed: one to supply air for combustion and the other to induce draught; the arrangement is referred to as a balanced draught system. This is the most sophisticated of all draught arrangements, as it permits very close control over combustion air supply and chimney draught conditions.

chimney draught – choice The particular choice of a draught system depends entirely on the design and construction of the particular incinerator, boiler, etc and associated equipment, as well as on the intended method of operation. The plant assembly should be designed to minimise loss of draught and pressure to economise in the electrical energy input to the fan(s).

chimney draught – forced Using forced draught, the fan supplies the air for combustion, overcomes the resistance of the boiler and chimney system and provides the required efflux velocity. In some plants, the fan pressure is calculated to overcome the resistance of the boiler system only, the chimney draught providing the necessary pull for removal of the flue gases to the atmosphere. (See figure 20.)

chimney draught – induced A fan is inserted between the equipment flue outlet and the chimney. Usually, the fan is located in a by-pass to the main flue connection, the latter being fitted with a damper which is closed when the induced draught fan operates. The discharge pressure of the fan must be sufficient to

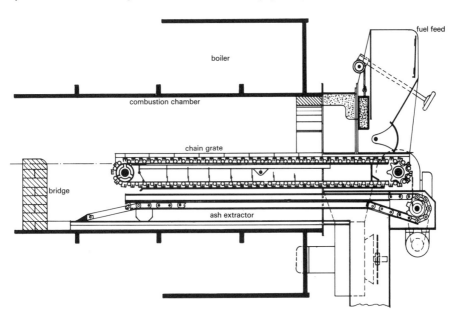

Figure 18 Chain grate stoker

discharge the flue gases against the resistance of the chimney circuit and to provide the required efflux velocity. The fan suction pressure must be adequate to overcome the friction through the incinerator, the boiler and the firing equipment. (See figure 21.)

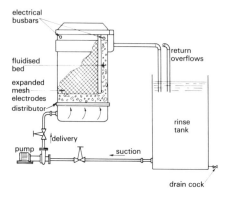

Figure 19 Diagrammatic arrangement of a Chemelec cell application

chimney draught – mechanical Relies on use of fan(s) to motivate the flue gas evacuation. Mechanical draught operation permits control of the chimney function regardless of weather and boiler conditions. Permits use of smaller size chimneys; can be aid to energy conservation.

chimney draught – natural The chimney operates under the natural buoyancy forces caused by the difference in densities between base and terminal due to the temperature differential between the flue gases and the ambient air. (See figure 20.)

chimney draught – secondary and primary air supply The air supply for combustion is made up of the theoretical minimum air quantity calculated to give combustion of the fuel (the primary air) and of the excess air required in practice to burn the fuel completely (the secondary air). When the combustion air is supplied by a forced draught fan, then either all the air may be delivered in one stream, or it may be divided into two separate air circuits supplying the primary and the secondary air respectively. The *primary air* is commonly delivered from underneath the fuel bed via the ash pit, or directly around the oil or gas jet when such fuel is employed. The *secondary* air would be delivered above the solid fuel bed or around the outside of the cone of primary air and atomised oil. The adoption of such a split

system of air supply permits good control over the final stages of combustion and over smoke emission.

chimney efflux velocity (exit velocity) The velocity at which the gases leave the chimney terminal. This is modified by the specific design for minimum cross-sectional area of the flue gases. If the efflux velocity is too low, the plume of gas leaving the chimney terminal tends to flow down the outside of the stack on the leeward side; the effective chimney height is thereby effectively reduced. The maintenance of a high efflux velocity will avoid this downwash. This velocity should be about 6m/sec for natural-draught operation and between 7.5 to 15m/sec for mechanical-draught systems. Incinerators equipped with forced-draught fans only should have a chimney efflux velocity of not less than 6m/sec when operating at full output. Incinerators equipped with induced-draught fans should have a chimney efflux velocity of not less than 7.5m/sec at full load for outputs up to 9,000 kW, increasing to a maximum of 15m/sec at full load for outputs of 135,000 kW. (See figure 23.)

chimney fitting Any flue-gas-carrying component other than a chimney section, eg chimney bend, chimney tee or terminal.

chimney section A straight component of uniform bore (except for any swage, socket or spigot) comprising flue lining, thermal insulation and outer casing.

chimney support An accessory designed to support the load of the chimney. (See figure 24.)

CIBS Abbreviation for Chartered Institution of Building Services Engineers.

circuit-breaker Electric switch designed to make and break a circuit under normal conditions; also to break the circuit under abnormal conditions of overload or short circuit.

circulating pump A device (usually electrically or – more rarely – diesel-motor driven) used to move or circulate a liquid.

Cistermiser Trade name of a device for reducing water wastage in the use of automatic flushing systems in mens' toilets. Non-electric mechanical control valve which automatically shuts down the water supply to such systems when they are not in use. Offers considerable

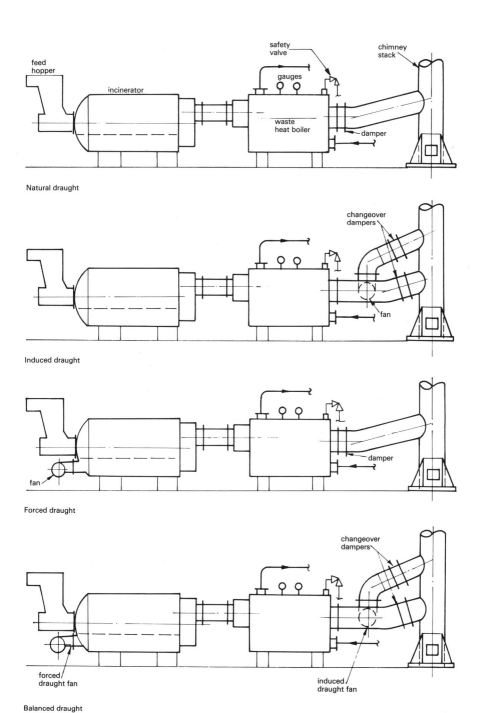

Natural draught

Induced draught

Forced draught

Balanced draught

Figure 20 Boiler draught conditions

41

scope for reducing water consumption.

Clark's Process Method of softening hard water by adding lime to react with the free carbonic acid and the carbonic acid combined in the bicarbonates of calcium and magnesium to form insoluble carbonates and hydroxides which precipitate.

Clean Air Act (1956) – UK Scope includes control of dark smoke, smoke from furnaces, smoke control areas, special cases of smoke emission. Its object is to reduce or eliminate the emission of dark (objectionable) smoke within smoke control areas. Associated with Memorandum on Chimney Heights, which gives directions as to minimum heights of boiler and furnace chimneys in certain categories of areas, and lays down acceptable chimney efflux velocities.

clean room An area designated for a high standard of air filtration. Generally uses high-efficiency particulate air filters capable of removing 99.97 per cent of all particles in the supply air of 0.3 micron diameter or more. Air distribution commonly achieved by downward displacement of the air from a plenum above a ventilated (perforated) suspended ceiling. *Application:* electronics, pharmaceutical, microbiology and similar industries.

clearness The atmosphere is not usually as clear at sea level as at, say, 1 mile (1.6km) above sea level; thus the radiation level, or insolation, is not as valuable. This phenomenon is defined by 'clearness' numbers. In parts of India and in the Arizona desert, clearness numbers approach 1.0.

clearness factor States the natural clearness of the atmosphere in a particular location; varies over the season of the year. Is applied to clear-day insolation value to establish actual insolation.

clearness number See *Clearness factor*.

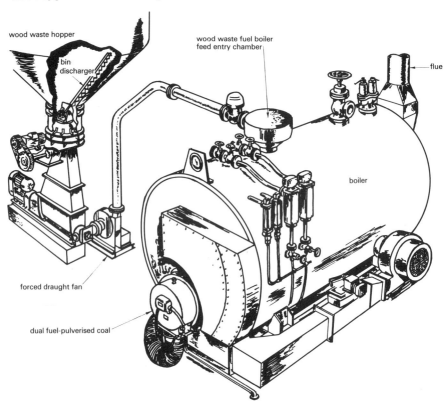

Figure 21 Induced draught boiler

clinker Porous mass formed when particles of hot ash are subject to surface melting and thereby fuse with the adjacent particles.

closed-cycle gas turbine Operates somewhat similarly to a steam turbine. Compresses and heats the gas, does work in the turbine and recycles within a closed cycle. Working fluid must be a near-perfect gas, eg helium. (See figure 25.)
Application: restricted by very high cost.

coagulant A substance introduced to precipitate suspended solids in water, usually a scale of aluminium or of iron. The precipitate (or floc) is basically a hydroxide.

coagulant aid A substance which, when added to water containing a coagulant, intensifies the settling action and appears to make the floc denser.

coal – brown Commercial term for lignite – a coal of relatively low energy content – in order of 14MJ/kg – available in large deposits of wide quality variations.

coal – hard Grade of coal with high energy content – about 60 per cent carbon by weight.

coefficient of performance (COP) Used in relation to heat pumps or refrigeration machines. It is the ratio of the amount of heat delivered to the energy input to the machine. Its value lies commonly between 3 and 6 for compression refrigerators and less than 1 for absorption machines.

cold draw Related to the design of bellows-compensated pipe systems. Allowance for the *cold draw* is made in the layout of the piping by making a gap between the flanges of the pipe section adjacent to the bellows joint equal to the free length of the fitting (as measured across the flanges) plus a length allowance of half the traverse listed for the particular joint. When the flange bolts are subsequently drawn up and tightened, the bellows joint will have been extended by the required amount. Certain bellows type joints designed for operation at high pressures do not permit cold draw; the total movement is then taken in compression from the free length of the expansion joint.

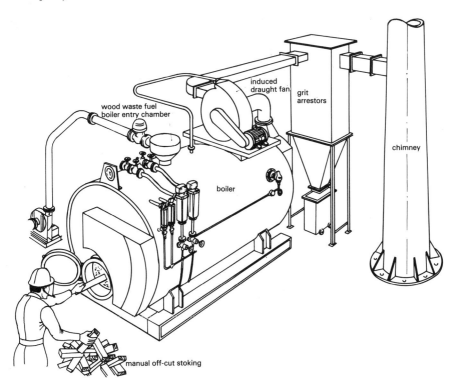

Figure 22 Alternative arrangements for furnace draught provision

coke Solid fuel obtained by the heat treatment of suitable coal; commonly a by-product of gas production (town gas). Used to be widely employed in space heating boiler plant; industrial users are steel works which require this fuel for smelting, reduction, etc.

Coke is a high-energy fuel, containing about 80 per cent carbon-calorific value 28MJ/kg. Has relatively high ash content.

cold drawn tube Mild steel pipe (tubes) which

has been drawn cold through a die and over a mandrel and thoroughly annealed. The cold-finishing process produces tubes to finer limits than possible where the manufacture ceases at the hot-finished stage.

cold spring Allowance for thermal expansion in the design and layout of a piping system. The expansive movement is computed from the room temperature to the operating temperature and all runs of piping are made somewhat

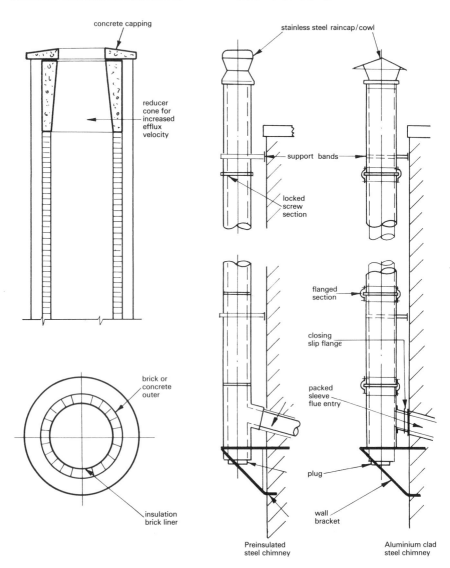

Figure 23 Chimney terminals for mild steel masonry and pre-insulated packaged chimneys

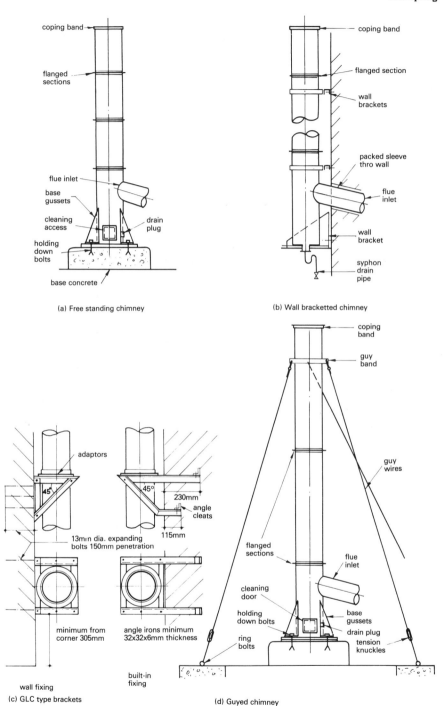

Figure 24 Various types of chimney supports

Figure 24 Various types of chimney supports

collection efficiency

shorter than the measured cold dimensions. The pipe, having been cut short by one half or more of the calculated expansion length and then sprung into place, will assist during assembly of the system and also tends to control the distribution of stresses and reactions betwen the hot and cold conditions, particularly where 'creep' is not a material factor.

Pipe systems which operate a high temperature where creep tends to make the pipeline relax into the hot configuration are subject to different considerations.

collection efficiency Ratio of the usable solar energy supplied by the system to the solar energy incident on the collector surface over a stated period (instantaneous, daily, annual).

collector angle Relates to solar collectors. The angle between the collector and the horizontal.

colloids Suspended matter in a particularly finely divided state. The dividing line between coarsely dispersed material and colloids is where the particles are just visible under an ordinary microscope, having diameters of less than about 1um. Colloids carry small negative electric charges, which in relation to their weight are large enough for the particles to repel each other and remain in suspension.

colo(u)r temperature The temperature, in degrees K, at which the shape of the spectral irradiance of a full black body radiator most closely matches the shape of the spectral irrad-iance curve of the radiant source under consid-eration (eg the colour temperature for global solar radiation over the wave band 300nm to 780nm at the earth's surface in the United Kingdom is typically around 5,700°K, with substantial variation).

column radiator See *Radiator – column.*

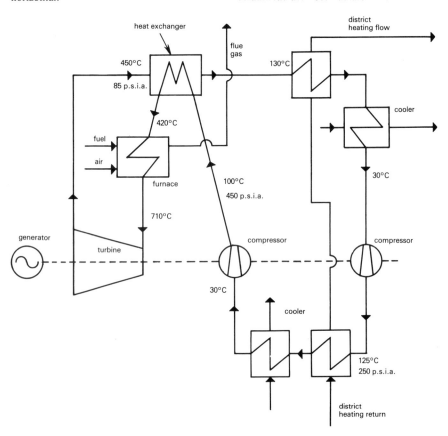

Figure 25 Diagrammatic representation of a closed-cycle gas turbine

combination boiler Includes within one casing the heating boiler and the associated hot-water indirect cylinder (calorifier). Usually, the latter is of the rapid-recovery type and thus smaller in size than a conventional hot-water indirect cylinder for the same hot-water supply duty. Boiler can provide a supply of hot water rapidly. System can be arranged to give first priority to meeting the hot-water demand before providing space heating.
Advantage: compact arrangement occupying minimum floor space.
Disadvantage: possible difficulties in controlling the hot-water supply temperature; correct forecasting of hot-water requirement (ie size of hot-water rapid-recovery cylinder).

combined heat and power system (CHP) Generation of heat during production of electric power and utilisation of the heat in a district heating system.

combined heat and power turbine system Gas turbine providing shaftpower for driving electrical generators, compressors or pumps and simultaneous heat in the form of hot water or steam from a single fuel input.
Application: the systems find wide-scale application in industry, particularly where a continuous and steady heat demand relative to electrical power is a major contribution to production services. They are also appropriate to those commercial enterprises where sustained heat demand and fairly low power demand exist.
Energy-saving potential: it has been demonstrated that under optimum conditions heat can be produced at an efficiency of 80 per cent on gross cv and simultaneously power can be produced at 90 per cent efficiency.

combined light and ventilation fitting Permits the supply or extraction of ventilation air at the light fittings; more commonly used for extract purposes to withdraw the heat generated at the light fitting before this enters the ventilated (air-conditioned) space. Has effect of cooling light fitting, thereby extending its life. Aid to energy recovery schemes. Reduces number of fittings inserted into ceiling. Relatively expensive, depending on type used. Can be modular.
Application: air-conditioning schemes with energy recovery provision. Integrated ceilings.

combustion Chemical reaction which generates heat to take place between the fuel (combustible material, commonly carbon, hydrogen or compounds of same) and oxygen in the air in such a way that the heat output is maximised.

combustion – spontaneous Heating of coal in storage noticed as a fire in the coal store (stack). The critical temperature in the storage of coal, at which such combustion is likely to occur, varies with circumstances of storage and type of coal; likely to be within range of 55°C to 70°C (131°F to 168°F), though lower critical temperatures have been recorded.
Most likely to occur between four weeks and four months of stacking the coal. After eight months, the risk is likely to have passed.
Large stacks must incorporate means of monitoring the stack temperature for early warning of excessive heating within the store to permit remedial action to be taken before there is a fire (spontaneous combustion) in the coal store.
Correct methods of stacking and ventilating large coal stores are known and should be adopted.

combustion air requirement Quantity of air which must be continuously supplied to support and maintain combustion of a material; can be theoretically calculated, based upon element molecular weight principles, which yield the following statements:

12kg of carbon plus 32kg of oxygen produce 44kg of carbon dioxide

4kg of hydrogen plus 32kg of oxygen produce 36kg of sulphur dioxide.

Air contains only 23 per cent oxygen. The *theoretical* air supply quantity for the complete combustion of 1kg of fuel containing these three combustible elements can be mathematically stated as follows:

$$\frac{1}{0.23} \left(\frac{32}{12} C + 8H + 1S\right) kg$$

C, H and S are the carbon, hydrogen and sulphur fractional constituents of 1kg of the fuel.

combustion catalyst A chemical formulated to improve combustion properties (mainly of fuel oil) and to reduce fouling of the boiler flueways.

combustion – complete Of a fuel (or other material) in the presence of adequate air supply. The nitrogen in the air takes no part in the process. The combustible substances must reach a temperature at least equal to their individual ignition temperatures before they can unite with the oxygen for complete combustion.

47

combustion efficiency See *Boiler – combustion efficiency.*

combustion – excess air See *Excess air for combustion.*

combustion – fluidised See *Fluidised combustion.*

combustion – incomplete Usually associated with an inadequate air supply to the furnace and occurs when the carbon component is burnt to carbon monoxide instead of to carbon dioxide.
The analysis of the carbon dioxide (CO_2) and of the carbon monoxide (CO) in the flue gases of a conventional combustion process indicates the *quality* of combustion. Incomplete combustion is wasteful as regards the fuel, is likely to foul the flue gas passages and will result in smoke emission.

combustion – submerged See *Submerged combustion.*

combustion test kit – portable Portable combustion-testing equipment for measuring carbon dioxide, oxygen, carbon monoxide, smoke, draught, flue temperature and gas pressure supplied individually or in kits. A combustion analyser can give digital display for oxygen reading and combustibles or temperature.
Application: general use in industry and commerce.
Energy-saving potential: savings vary from about 5 per cent in new plants up to 40 per cent of fuel costs in poorly maintained, old plants.

combustion trim unit Microprocessor-based oxygen trim unit incorporating on-line combustion efficiency display and plant/transducer alarm monitoring. Capable of storage of two combustion characteristics for dual fuel firing.
Application: package steam and hot-water boiler combustion control. Industrial furnace combustion control.
Energy-saving potential: equipment payback period is claimed at less than 18 months on heat loads exceeding, say, 5,000kg steam/hour.

comfort zone Relates to environmental conditions created by space heating and/or air conditioning. Defined as that *range* of temperatures (or other environmental condition) in which the larger proportion of persons (office, factory, etc occupants) express themselves comfortable. For example, Dr Bedford (a pioneer in this field) found that, in terms of

dry-bulb temperature, the winter comfort zone for operatives performing light factory work ranges from 15.6°C to 20°C (60°F to 68°F) in the UK.

commissioning Adjusting the various control parameters of a (new) installation to meet the design performance. Essential activity before taking plant into general use.

compactor A precision-built item of machinery which incorporates an electrically operated compaction ram. It is designed to reduce a volume of waste matter by some 80 per cent and to package it into a clean and conveniently handled unit.

complete combustion See *Combustion – complete.*

composting (Organic) end product of refuse-recycling plant formed in a process of aerobic fermentation.

compressed air leakages Commonly occur at the air receiver mountings, hose unions, pipe connections, drain points, process equipment, quick-release couplings, etc. The power loss occasioned by such leakages can be material and losses of 30 per cent of total requirement have been established on test in some more or less typical compressed air systems.
Such scale of loss wastes expensive power and requires oversized plant. The following table indicates the air loss which occurs from individual leaks which are equivalent in area to 3mm and 6mm diameter holes in the pressurised system.

Leakage Air Loss from Compressed Air – m^3/min

Hole Size	Air Pressure (bar-g)					
(mm)	1.0	2.0	3.0	4.0	5.0	6.0
3	0.07	0.14	0.4	0.5	0.6	0.7
6	0.6	1.06	1.37	1.8	2.2	2.6

The wastage of compressed air can be converted to wastage of electric power by the following multipliers (m^3/min to kW):

Air Pressure (bar-g)	Single stage compressor	Two stage compressor
3	3.8	3.4
5	5.3	4.5
7	6.7	5.9
10	8.3	6.7

One method of evaluating the quantity of air

leakage is to operate the air compressor with all the air-operated equipment shut-off until the compressed air system reaches the full line pressure when the compressor will automatic-ally unload. A note is then taken of the length of time which elapses before the compressor comes back on load and for how long it runs before again unloading. As usual with such measurements, a number should be taken and averaged.

The total leakage (m³/min of free air)

$$= \frac{Q \times T}{(T + t)}$$

Q denotes the free air delivery capacity of the compressor – m³ of free air/min.
T denotes the time *on* load – minutes
t denotes the time *off* load – minutes
The power loss is then obtained by use of the multiplier.

compression pipe fitting Joints for (copper or stainless steel) pipes secured by means of a union and olive arrangement. The olive is slipped on to the end of the pipe, and the union nut tightens the olive on to the pipe when the nut is screwed to the fitting.

compression ratio Relates to refrigeration (vapour) compressors.

$$= \frac{\text{Head pressure of compressor}}{\text{Suction pressure of compressor}}$$

using same units for both pressures.

compressor unit Basically a single package comprising the compressor and drive with associated controls and connecting pipes to install at site to remote air-cooled condenser and to evaporator.

computerised optimiser A computerised optimiser using micro-electronic components, including a microprocessor.
Application: time/temperature control of industrial heating and air-conditioning install-ations.
Energy-saving potential: savings of up to 40 per cent on fuel can be obtained, the amount of saving being dependent upon the extent of controls previously installed and utilised.

condensate meter Measures the consumption of steam by the weight of condensate dis-charged from the consumer system. Figure 26 illustrates a typical rotating-drum condensate meter.

condensate recovery unit A packaged unit consisting of a receiver, one or two close-coupled pumps and all interconnecting pipe-work and valves.
Application: in plants using steam which is reusable as condensate.
Energy-saving potential: recovering condensate produces savings in fuel, water treatment and effluent. Total costs plus installation costs recoverabie within 12 months.

condensation Occurs when moisture-laden air comes into contact with a cold surface which is at a temperature below the dew point of the air. Water then separates from the air at the cold surface. (See figure 27.)

condenser Heat exchanger in which a vapour (eg steam, refrigerant gas) is condensed to a liquid, giving up its latent heat in the process. Liberated heat may be used for heat recovery purposes.
Condenser may be water-cooled or air-cooled.

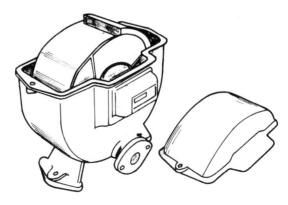

Figure 26 Condensate meter measuring supply of steam

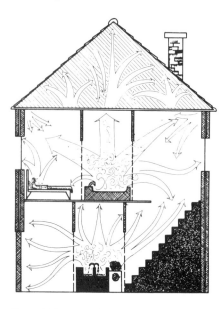

Figure 27 Sources of condensation in a typical domestic dwelling

condensing turbine See *Turbine – condensing.*

condensing unit Basically a single package comprising the compressor and drive assembly, the condenser (which may be air- or water-cooled), associated controls and all necessary connecting pipes. It will be connected at site to the evaporator of the system, and all components must be correctly matched for the specified conditions.

conduction Transfer of heat within a (solid) material by passage of the heat from the hotter portion through the layers of the material towards the colder portion of the material.

connecting piece An accessory for joining sections, or a section and a fitting, of the same nominal size. It is used only with sections and fittings without sockets or spigots.

constant flow valve A factory pre-set control device to give a constant fluid flow irrespective of variations in upstream and downstream conditions. Sizes from ½ in to 3 in with threaded ends and 3 in to 30 in flange mounted. Usually accurate to plus or minus 5 per cent. Operating range: 1 to 250psi and up to 230°C (446°F).
Application: for use in heating, ventilating and air-conditioning systems and in industrial

cooling systems.

contact factor (CF) Relates to cooler batteries in air-conditioning systems. Defined as the proportion of air passing through the coil that actually *touches* the cooling surface and is thereby cooled.

contactor Power-operated switch for making and breaking a circuit. Utilises the electric field generated by the contactor coil (connected to a low-power circuit, such as a sensing circuit) to activate the 'on' or 'off' action of a much larger power capacity.

continuous convector See *Convector – continuous.*

control (devices) The successful manipulation of an energy-using system requires controls to avoid overheating, boiling or frost damage and to achieve diversion into and out of the heat store. Such control devices are usually automatic and are pneumatically, electrically or thermo-mechanically actuated.

control level – float Actuates controlled equipment by change in level of a fluid which causes movement of the float.

convection Transfer of heat from one place to another by actual movement of the hot body. In fluids, the portions adjacent to the heat source become hot and expand, accompanied by a lowering of density. Motion is then set up under the action of gravity, with the lower density liquid moving away from the heat source, permitting the colder, denser liquid to approach to become heated in turn. Thus, heat transfer by convection takes place by reason of the movement within the volume of fluid. This process is also known as thermosyphonic.

convective radiator See *Radiator – convective.*

convector Heat emitter which gives out its heat predominantly by convection, though there will generally be also an element of radiant heat output from the heater surfaces.

convector – continuous Form of heater which provides continuous heating at the perimeter of a building; usually located under line of windows.
Can be arranged with partition and sound baffles to suit partitioning modules and layouts. May incorporate manually operated dampers to reduce the convection effect; the closer to the heater element, the more effective

the dampering.

convector – forced-draught Convector embodies a fan or fans and a compact heater battery. The fan propels the air over the heater to achieve rapid heat exchange. Greater heat output capacity than obtainable by natural draught. Amenable to thermostatic control by switching the fan on and off to vary the heat output.

convector – natural The convection effect relies on the natural up-draught through the heater (chimney-effect); the taller the convector, the greater this effect.

cooling – direct-expansion See *Direct-expansion cooling.*

cooling – evaporative See *Evaporative cooling.*

cooling pond Functions with water stored in an open pond. Cooling progresses by contact between the pond surface and the air passing over it, causing evaporation of some of the water. Usually, the cooled water is drawn off at one end of the pond and recirculated at the other end. Suffers from low rate of heat transfer, likelihood of contamination, requirement for large area and uncertainty of specific performance. Useful where ornamental pond required. Can also act as heat sink for an associated heat-pump system.

cooling tower A general classification of equipment designed for the evaporative cooling of water by passage of air through the equipment.
May operate with natural draught or be fan assisted (induced or forced). The water being cooled may be in *direct* contact with the air flow or it may be cooled *indirectly* within a closed circuit.

cooling tower – approach The temperature differential between the water leaving (off) the tower and the design (specified) air inlet wet-bulb temperature; eg for a water outlet at 27°C (80.6°F) and a wet-bulb air inlet at 20°C (68°F) the approach is 7°C (12.6°F).

cooling tower – bleed The quantity of water which is removed from the evaporative water circulation to control the content of dissolved solids in that water. Object is to prevent an excessive build-up of dissolved solids and/or of other possibly harmful atmospheric constituents. Bleed may be by intermittent or by continuous drainage; by manual or automatic activation to suit a water-treatment schedule.

cooling tower – blowdown As bleed.

cooling tower – carry-over The spray which is picked up by the air as it flows through the tower and which it carries out of the tower. Excessive carry-over can constitute a nuisance.

cooling tower – casing The envelope which encompasses the packing and the water system within the tower.

cooling tower – cladding As casing.

cooling tower – counterflow The fan (usually induced-draught) induces vertical air flow up the tower and across the packing in the opposite direction to the flow of water streaming down the tower, so that the coldest water is in contact with the driest air. Provides best cooling tower performance (effectiveness).

cooling tower – cross draught The fan creates a horizontal air flow as the water streams across the tower (usually operates with induced-draught fan). Permits the design of cooling tower having a low silhouette (which may be required under certain circumstances of the site). Must be protected against the recirculation of saturated vapour which is likely to occur in confined locations.

cooling tower – direct The circulating system cooling water comes into direct contact with the cooling air. Achieves optimum evaporative cooling, but inevitably causes contamination by atmospheric pollutants of the water which circulates through the cooled equipment. Unsuitable for installations where foreign matter (even though strained and filtered) is likely to choke small waterways, valves, etc.

cooling tower – distribution header Directs the water flow into the tower for best evaporative effect.

cooling tower – dosing pot Apparatus through which water treatment chemicals are introduced into the pipe system. Usually incorporates a metering device to achieve the specified rate of dosing.

cooling tower – drift As carry-over.

cooling tower – fill As packing.

cooling tower – forced draught The air is forced into the tower by the action of a fan which is fitted to the air inlet of the tower.

cooling tower – frost protection Means of

providing protection to obviate freezing of the water in the tower (and coils). Usually by electric immersion heater(s) or through by-passing some boiler heat through the tower in cold weather under thermostatic control.

cooling tower – indirect The water which circulates through the primary (cooled equipment) circuit is kept out of contact with the air by circulating the water through pipes over which the cooling air is directed. Protects the cooled equipment from contamination. Of larger dimensions than direct cooling tower. Cooling coils must be protected to avoid frost damage.

cooling tower – induced draught The air is drawn into the tower by the action of a fan which is fitted to the air outlet of the tower.

cooling tower – make-up The supply of water required to compensate for losses of water from the system through evaporation, bleed, pump glands or overflow. On the larger installations, make-up is supplied from a treated water reservoir; smaller installations may draw directly off the water main, possibly incorporating a water treatment dosing pot. Make-up water flow is usually controlled by a float and ball valve arrangement associated with the tower.

cooling tower – natural draught (atmospheric) The cooling air is drawn over the tower by the chimney effect of the arrangement. *Drawback:* depends on wind conditions, external temperatures, freedom from obstructions, etc.

cooling tower – packing The material inside the tower which functions as the heat-transfer surface over which the water is directed and distributed in its passage through the tower. May be timber, glass fibre or plastic.

cooling tower – purge As bleed.

cooling tower – range The temperature differential between the circulating water entering the tower and leaving the tower; eg if water enters at 40°C (104°F) and leaves at 30°C (86°F), the range is 10°C (18°F).

cooling tower – water treatment Required/provided to obviate fouling of the cooling tower system through the growth of algae, scale or/and corrosion.

corner-tube boiler Consists basically of an all-welded cage system of corner tubes with interconnecting longitudinal and transverse headers, associated with welded-in heating surfaces such as furnaces walls and convection heat banks. The corner tubes ensure that the cooling system is self supporting. For the smaller size of boiler, the furnace is made completely gas tight by being enclosed in a sheet-metal casing which moves with the expansion and contraction of the tubes. the larger size boilers employ integral fins on the furnace tubes.

A corner-tube boiler can be designed to function as either a hot-water or a steam boiler or as a combination of both.

Boiler used for the larger group heating schemes and for district heating; can be supplied in the form of a fully packaged boiler house (see *Packaged boilers*) and in this form has a rating of between 5 and 10 MW. The maximum rating for a corner tube boiler is in the order of 160 MW.

One particular advantage of the boiler is the relatively low weight: typically for a 10 MW unit is 11 tonnes empty – 15 tonnes when filled with water (less than half the equivalent weight of a three-pass boiler).

As a special feature, it can be designed to operate as a combined steam and hot-water boiler. One popular model provides hot water and steam in a 70:30 ratio. In the steam model, the steam drum is mounted directly above the boiler. One interesting application is to use the steam component generated in a predominantly hot-water boiler to provide the steam for soot blowing, oil heating, oil jet atomising, etc. The boiler contains no refractories, apart from the oil burner quarl, thereby saving weight and maintenance costs. A gas-tight membrane avoids low temperature corrosion problems within the furnace.

Applications: portable hot-water boiler house, say, for use during the initial stages of group-heating schemes; large permanent hot-water boilers for major group heating systems; hot-water boilers for stand-by and peak duties in large district-heating schemes; waste heat recovery in conjunction with municipal refuse incinerators; waste heat extraction from the flue gases of gas turbines; stand-by steam generators for medium-pressure steam plant of pressures up to 105 bars at 520°C (968°F). High boiler efficiencies (up to about 90 per cent) are claimed for this boiler type. The boilers are suitable for firing with coal, gas, oil, peat and certain types of wastes with a turn-down ratio to 30 per cent.

Corner-tube boilers are the most widely used boiler type in European group- and district-heating schemes.

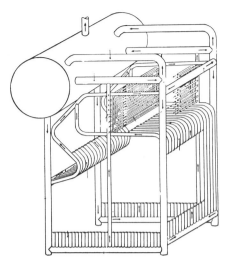

Figure 28 Schematic arrangement of a corner tube boiler

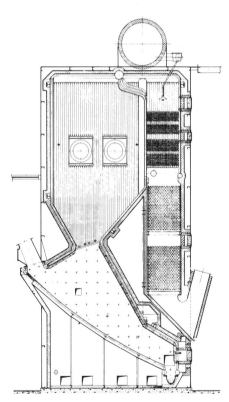

Figure 29 Sections through a smaller corner tube boiler

Cornish boiler Brick-set steam boiler with single internal flue shell. Used to be popular in the coal burning era when efficiency not paramount. Sizes to 2,000kg steam/hour and 2m dia. x 8m long. Easy to maintain by low-grade labour.

corrosion Usually causes 'pitting' or chemical attack on equipment and/or pipe surfaces. Consequences of corrosion may be serious.

corrosion – aerobic Iron bacteria, present in some waters, are *aerobic*. They absorb oxygen and oxidise any ferrous components they meet, whether in solution in the water or from the pipe metal itself. If the iron content of the water being conveyed is high, slimes are formed which affect the potability of the water.

corrosion – bacterial anaerobic Occurs in presence of sulphate-reducing bacteria; these are particularly active in their effect on iron and steel, but do not attack same directly. Reduce sulphates to form hydrogen sulphide, which combines with iron to form graphite. Present in water: sulphuric acid results which then attacks the metal externally or internally, depending on whether bacteria presence is in soil or water. Resultant pitting is soft and can be dug out with penknife. Corrosion pits have a characteristic circular shape; irregular areas of pitting are due to a number of circles merging. Virulence of attack caused by continuous process, the bacteria feeding on by-products of the corrosion products created by their activity.
Bacteria are *anaerobic* and cannot exist in presence of oxygen. Difficult to detect; samples must be taken in bottles which have been sterilised and filled with an inert gas or by other specialised methods.
Cause: generally, poor drainage.
Avoidance: pipes should not be laid in wet areas where there is no access to oxygen; alternatively, pipe runs should be drained to let in air; where this is impracticable, special pipe coatings and/or cathodic protection should be provided.

corrosion – electrical See *electrical corrosion*.

corrosion – fretting See *Fretting corrosion*.

corrosion inhibitor Chemical introduced into a piped system to inhibit internal corrosion, eg some forms of corrosion generate hydrogen gas within a heating system which causes the symptoms of air locking; inhibitors available to combat this.

counterflow shell and tube heat exchanger
Shell and tube counterflow heat exchangers of
U-tube and spiral tube configurations.
Applications: heat recovery from the cooling
of liquids with applications particularly suited
to high-temperature/high-pressure duties with
disparate flows including thermal fluid
applications. Also for use as calorifiers for the
direct conversion of superheated pass out or
back pressure steam.

CPU Relates to expansion joints.
Abbreviation for 'total axial traverse from
cold-pull-up'.

crawlway Service duct of dimensions to permit
a normal sized person to crawl through.
Term may also be loosely applied to larger size
service duct. (See figure 30.)

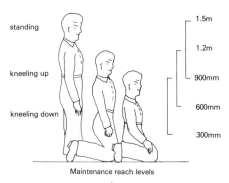

*Figure 30 Dimensions of accessible manholes
and inspection pits*

crystallisation Related to absorption refrigeration equipment. If the lithium bromide
solution becomes too concentrated, it changes
from liquid to solid form (crystals). The maximum possible concentration decreases as the
solution temperature decreases. If a solution is
close to its maximum concentration and if its
temperature is then lowered, it will crystallise.
Can be a serious problem in absorption refrigeration systems, as the crystallised lithium
bromide will then block the pipes, and plant
operation breaks down. Three major factors
can be responsible for causing an unwanted
drop in the temperature of the solution: power
failure; air leakage into the system;
maintenance of a condensing temperature
which is too low for the system.
Once crystallisation has occurred, it is reversed
by heating the pipe at the blockage(s). Auto-

matic de-crystallisation means can be incorporated with the system.

cubical expansion (specific) The increase in
volume which unit volume undergoes when its
temperature is raised through 1 degree.

curing Term used in connection with
refractories; relates to the time required to
permit settling down of cast refractories before
the application of heat. In no circumstances
should the curing time be less than 24 hours; it
may well be longer, depending on the specification of the refractories used.

cycle The complete sequence of values of a
periodic quantity which occur during a period.

**cyclone furnace – incineration/waste-heat
recovery** Cyclonic system of combustion
burning waste products to produce heat from
resultant waste gases as steam, hot water or air
via a boiler or heat exchanger. Incineration
occurs in a separate chamber, and in many
instances existing boilers can be utilised. (See
figure 31.)

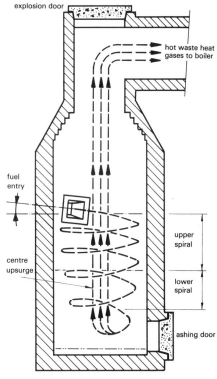

Figure 31 Cyclone furnace

Application: incineration and waste-heat recovery from wood waste, scrap tyres, sludges, etc (wet and dry materials).

cylinder jacket For industrial and domestic hot-water cylinders and tanks. Usually manufactured in wrap-round quilt with zip fasteners and vents to accommodate pipework.

cylinder and pipe thermostat Functions with

changeover snap-action switch supplied with fixing strap for cylinder or pipe mounting. *Application:* to sense hot-water temperature and switch off boiler or pump when the required temperature is reached.

cyrogenics Science and technology of substances used at very low temperatures – say 250°C (418°F).

Dalton's Law The pressure which a vapour exerts in a mixture is very nearly independent of the pressure exerted by any other gas or vapour in the mixture.

damper – butterfly See *Butterfly damper.*

damper – fire See *Fire damper.*

damper – iris See *Iris damper.*

dawson joint An all-welded pipe joint in which the weld is built up between the pipe ends and is carried down into an internal nipple; forms an extremely rigid joint.

dead leg Applies to hot-water supply systems and denotes any length of pipe within the dis-

tribution system which is not part of the hot-water loop (circulation); eg final connection to a draw-off tap.

Some water supply companies limit the permissible length of a dead leg to reduce work of water.

deaerator Reduces the oxygen concentration in boiler feed water. Mechanical deaeration methods for distilling water can reduce the oxygen concentration to 0.1ppm. This is not adequate for high-pressure boilers, for which vapour purification, flash-type deaerators or desorption methods are employed which can reduce the oxygen concentration to virtually 0, with the addition of hydrazine dosing. (See figure 32.)

D

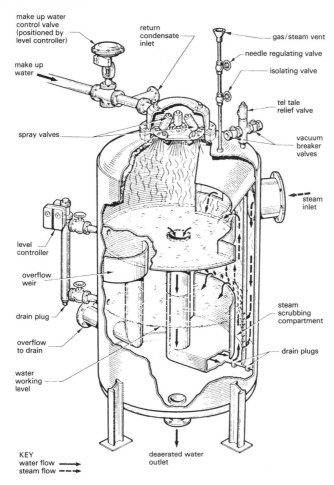

Figure 32 Section through an industrial water deaerator

declination The tilt of the axis of the earth relative to the sun. Measured by the angle created by the noonday sun and the equatorial plane of the earth. Varies from $+23°\,27'$ to $-23°\,27'$.

de-frosting Essential with air-cooled evaporators which operate at temperatures below the freezing point (0°C; 32°F). Accumulated ice deposition will obstruct heat transfer and *must* be periodically removed. Manual removal of ice deposit is seldom a realistic proposition. Practicable methods in common use are:
   By electricity; electric resistance elements are placed within the coil or directly below;
   By hot gas; pumped from the compressor; this method assumes that there are multiple evaporators which are defrosted in turn, as the compressor must be working to supply the hot gas;
   Reverse cycle; in which the direction of operation is reversed, so that the evaporator is made to act in the condenser mode; another evaporator or heat storage is required as source of heat for this method;
   By water spray; which floods the coil periodically to melt the ice. Drip trays and drain pipes must be provided to discharge the liquid.
Most systems incorporate automatic defrosting controls and timers.

degree-days method (essentially a climatic parameter) Means of comparing, over different periods, the variations in the seasonal heat loads of heating installations in different geographical locations. Assesses for monthly periods the daily difference in temperature (°C) between a base of 15.5°C and the 24-hour mean external temperature. The monthly totals are then used to compare monthly changes in the weather factor, or they may be added together for the complete heating season to permit comparison between one year and another and one location and another in regard to the severity and duration of the winter.

dehumidification Process of abstracting moisture from air.

dehumidifier Apparatus for achieving dehumidification of air. Operating principle may rely on dehydrating agent (such as silica gel), or on cooling of air below its dew point to shed moisture.

delayed-action ball valve Permits water in a storage tank to be drawn down to a predeter-mined level before the valve opens, thereby avoiding the dribble condition at opening and closing which is otherwise likely to occur. Water-treatment plant installed in line to the storage tank therefore receives a full flow of water at all times; once the tank has been filled, the ball valve closes quickly and cleanly. The feature of the delayed-action ball valve which permits a predetermined amount of water to be drawn off before the valve opens and readmits make-up water can be utilised with boosted systems to avoid frequent starting and stopping of the booster pump(s) whenever a small quantity of water is drawn off the storage tank.
The operating principle of this ball valve: the ball floats within a separate canister, at the base of which a valve (v) is operated by the float beneath. As water is drawn from the tank, the ball valve remains shut, while the ball continues to float within the canister. When the water level has reached the required depth, the lower float opens the valve (v), empties the canister and causes the main ball valve to open. As the level rises, so does the bottom float, thus closing the valve (v), and hence the ball is not raised until water overflows the rim of the canister and then causes the main ball valve to quickly and sharply shut off.
Application: water supply to water-treatment equipment, boosted water to high-level storage, etc.

deodorising filter Activated carbon deodor-ising filter.
Applications: removes weld fumes, vehicle exhaust and engine burn-off, and improves garage and factory environment control.
Energy-saving potential: most installations are said to have recouped their installation costs within three years through fuel savings.

design condition(s) Parameters laid down for the execution of a particular services design; eg minimum external temperature (heating), max-imum external temperature (cooling), noise level, standard of air filtration.

de-superheater Relates to steam-heating systems in which the superheat of the steam must be removed before the steam enters the heating (process or comfort) equipment. With-out de-superheating, the superheated steam must be cooled within the heating equipment before it can give up its latent heat – this would prolong the heating process. De-superheating is effected by spraying controlled quantities of water into the steam in a superheater vessel.

dew point Temperature of the air at which the saturation vapour pressure equals the actual vapour pressure of the water vapour in the air. If moist air is cooled below its dew point, it must shed moisture to maintain the saturation vapour pressure at the new reduced temperature of the air.

dew point – acid See *Acid dew point.*

dezincification Corrosion of brass products, usually indicated by heavy white deposits on the brass wear that is in contact with water. Brass is an alloy of copper which contains zinc. Whilst zinc is widely used for protection against corrosion, it can itself be attacked by waters which contain free carbon dioxide or which have a high pH (in excess of 8.2) and a high ratio of chloride to carbonate hardness; in the latter, for a low chloride content, the problem arises when the ratio of chloride to carbonate hardness is greater than 1, and for chloride greater than 20mg/litre; dezincification can occur when the ratio is lower. *Remedy:* avoid hot-pressed brass and/or increase the carbonate hardness in applications where dezincification is likely or suspected.

diaphragm valve See *Valve – diaphragm.*

differential thermostat See *Thermostat – differential.*

diffuser See *Air supply diffuser.*

diffuse radiation See *Radiation – diffuse.*

digital temperature indicator Electronic thermometer providing accurate digital readout of temperature from thermocouple or resistance thermometer input.
Application: boiler temperature and environmental air temperature measurement.

dielectric heating In dielectric heating, the article to be heated is placed between metal plates or electrodes, which are connected to a source of alternating current. The article thus constitutes the dielectric of a condenser, and as the condenser is not perfect, losses take place in the dielectric which are converted into heat within the article.
The process of dielectric heating may be pictured by considering the action of the alternating supply on the molecules forming the material being heated. When the electric field is applied to the electrodes, molecules in their vicinity acquire electric charges, the polarity of these charges at any instant depending on that of the impressed voltage. A rapidly alternating voltage will reverse the polarity of the molecules a great number of times in every second, and the higher the supply frequency, the greater the number of reversals. These changes in polarity set the molecules into vibration, with the consequent generation of heat inside the material. The actual amount of heat so produced is a function of the frequency and of the 'loss factor' of the material considered. The 'loss factor' represents a convenient way of comparing the different rates of heating of different materials. As the energy dissipation by the vibrating electric field takes place throughout the material, the heat generated is distributed uniformly throughout the body of the material. (See figures 33 and 34.)
Applications: the glueing of plywood, curing of plastics, etc.

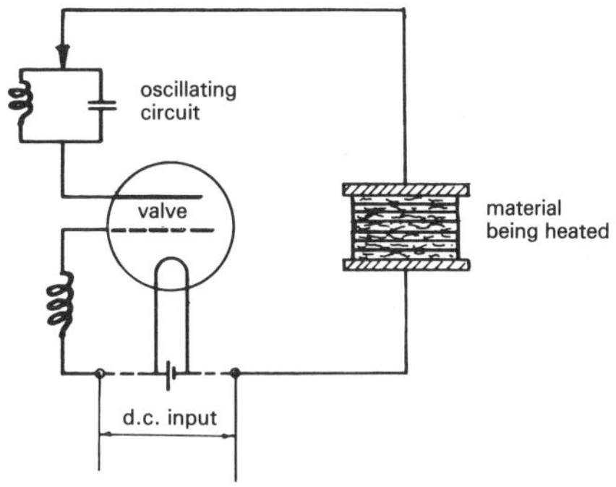

Figure 33 Principle of dielectric high frequency heating

diluted gas boiler flue system See *Fan-diluted gas boiler flue system.*

direct-acting controller Comprises a sensing element which, by the thermal movement (expansion or contraction) of a liquid or vapour, transmits an activating power *directly* to a bellows or diaphram, which then actuates the controlled device (eg a damper, motorized valve, pneumatic controller, etc).
Application: generally limited to the smaller installations where small independent control units are involved. Not suitable for multiple and/or central control.

direct-coupled centrifugal circulating pump Pump and electric motor are mounted on a common bed plate and are directly coupled by a shaft. Capable of large duties and commonly used on large-scale heating and industrial projects.

direct current

Large output pumps are provided with water-cooled bearings.
Pump tends to be noisy in operation. It lacks the flexibility for adjustment of performance of the belt-driven pump.

direct current Unidirectional electric current.

direct-expansion cooling Refrigerant evaporates in coils which are placed directly into the air stream being conditioned, thereby cooling the air. Arrangement difficult to control within close limits. (See figure 35.)

direct gain Relates to passive solar-heating systems in which the solar radiation is permitted to penetrate directly into the building via the windows and thereby provides space heating to the affected areas.

direct-on-line starter See *Electric motor starter – direct-on-line.*

direct radiation See *Radiation – direct.*

direct steam heating See *Steam injection.*

discharge lamp In this, light is produced by the passage of electricity through a metallic vapour or gas which is enclosed in a light bulb or tube.

distribution board (dis. board) Panel or board from which electrical connections are taken off for the distribution of electricity to various electrical circuits; usually embodies fuses or circuit breakers to protect the circuits served off the board.

district heating – single-pipe See *Single-pipe district heating.*

district-heating system Serves a large number (say, more than 1,000) consumers of heat which are independently connected to it, usually via a consumer terminal which includes a heat-metering facility. (See figure 36.)

district-heating system – diversity Relates to a system of consumption (electrical, water, steam, etc) which serves a number of consumers; each individual consumer is unlikely to always draw from the system the maximum quantity or units allocated to him in the design of his installation. The system is said to embody diversity; ie the total supply into the system at any one time will be less than the sum of the individual maximum allocations.

district-heating system – factor A coefficient of less than unity; applied to a supply system serving a group of individual consumers to allow for the assumed diversity of consumption. The smaller this factor, the smaller the size of the mains network into which the indiv-

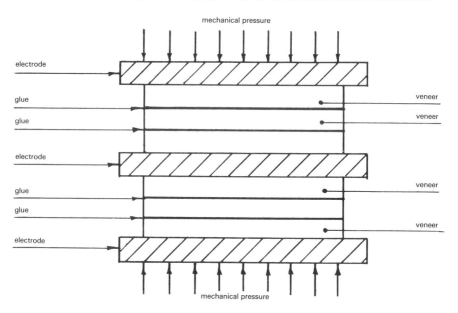

Figure 34 Application of dielectric heating to the gluing of wood veneer

60

idual consumers are connected.
Published recommendations and rules of thumb are available for the particular diversity factor to be adopted in given circumstances. The larger the network, the more important the correct assessment of the diversity factor.

double-inlet fan See *Fan – double-inlet.*

double-pole switch A cut-out, circuit breaker, fused switch or similar in which the circuit is broken at both poles simultaneously.

double-skin chimney Consists of an inner and an outer steel shell. The annular space between the skins may be filled with a suitable insulation material (such as mineral wool). The construction must avoid any direct contact between the inner and outer shells.

doublet Refers to geothermal schemes. The combination of a pair of geothermal wells, one of which produces hot water from the underground aquifer and the other reinjects the cooled water back into the aquifer; usually the two wells are spaced about 1 km (horizontal distance) apart.
The wells may *both* be drilled vertically or they may be inclined, so that the well-heads are closer together and minimise surface piping.

drainage – electrical See *Electrical drainage.*

draught diverter See *Gas diverter.*

draught excluder – brush See *Brush strip draught excluder.*

draught-proofing See *Weather-stripping.*

draught sealers Range of draught sealers designed to cope with most draught problems associated with doors and windows.
Application: for sealing internal and external doors and hinged windows.

draught stabilizer Weighted metal plate (or combined with access door) fitted into base of a chimney for the purpose of regulating the chimney draught by permitting the stabilizer to swing into various positions between closed and wide open and thereby draw relatively cool air into the chimney. Fitted with adjustable balance weight(s). Can also act as explosion relief door. Care necessary to prevent 'overcooling' of the chimney, which may lead to emission of smuts.

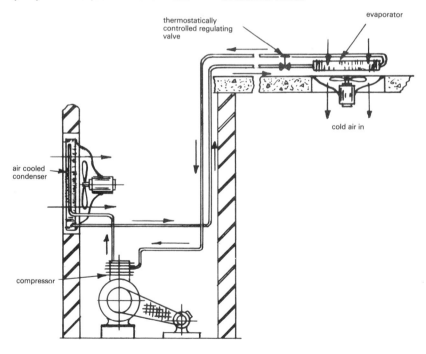

Figure 35 Basic direct-expansion refrigeration system

61

drum wall Relates to solar-heating systems. A thermal storage wall made by stacking large (say, 200-litre capacity) drums containing water one on top of the other; usually stacked horizontally and one drum deep.

dry-bulb temperature temperature detected by the dry bulb of a thermometer (as opposed to wet-bulb temperature).

dual gas burner Provides immediate fuel change from natural gas to butane (or propane) or vice versa, even when the burner is firing. Changing fuel according to price advantage, or where interruptable natural gas supplies exist, is possible. The different characteristics of natural gas and, for example, butane are dealt with economically by a basically conventional gas train incorporating two sets of gas pressure regulators, one for each gas. The two gas inlet sections are connected by short and joined levers in the form of a handle. The interlocked lever assembly operates the gas cocks so that as one opens the other closes. With dual-fuel gas and oil firing, the burner controls are necessarily different for each fuel, and the changeover from one fuel to another is less immediate than with dual gas firing.

duct(s) Formed conduit(s) to convey services or gases. May be constructed of masonry, concrete, steel, glass fibre, asbestos, etc.

ducting Arrangement (system) of ducts.

ducting – fabric Ducts constructed of fabric components interlocked mechanically with a supporting metal spiral. Commonly coated with butyl-coated cotton; for non-inflammable finish coating would be neoprene-coated glass.

ducting – spiral Ducting consisting of a wire helix spring-steel frame to provide flexibility. Available in thermally insulated form with glass fibre insulation and vinyl vapour barrier.

ductstat Detector with its detecting bulb located inside a duct, in order to activate motorised valve(s), damper motor(s), air controls to achieve control over temperature, pressure, humidity, etc.

dump condenser Dissipates excess steam by condensing it in a fan-assisted heat exchanger. Heated air usually dumped to atmosphere (wasteful); may be incorporated into energy-saving scheme.

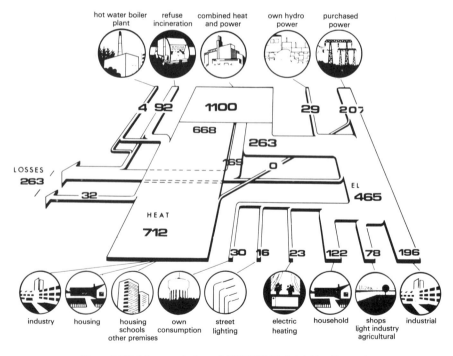

Figure 36 Total energy balance LINKOPING district heating system

dust collector Removes dust particles from an aerosol dispersion through displacement of the dust and its subsequent removal. May utilise gravity, centrifugal or electrical motivating forces. (See figure 37.)

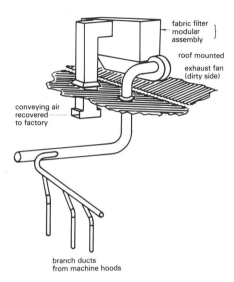

fabric filter modular assembly }

roof mounted

exhaust fan (dirty side)

conveying air recovered to factory

branch ducts from machine hoods

Figure 37 Waste extraction to fabric filter

dust collector – centrifugal Directs dust particles to describe a number of revolutions within the dust collection zone; the resulting centrifugal forces are utilised for dust collection. A well-known example is the cyclone. (See figure 38.)

dust collector – combined gravity and centrifugal Encourages centrifugal forces by single or multiple deflections situated within the collector; to assist the gravity forces in the dust collection.

dust collector – displacement velocity Velocity of the dust particles towards the removal zone relative to the velocity of the carrier medium.

dust collector – displacement zone Zone within the dust collector where the dust particles are dispersed out of the carrier medium in the direction of the removal interface.

dust collector – electrical (electro-precipitator) Utilises electrical forces to collect the dust from a carrier medium. Electric charges are introduced into the electrical field and exert a force on the charged particle in the direction of the potential gradient. (See also *electrostatic air filter*.)

dust collector – filtration Removes dust particles through their adhesion to a suitable filtration material (fabric, viscous liquid, granulated material).

dust collector – gravity Utilises the forces of gravity to displace the dust particles at the settling velocity towards the removal interface.

dust collctor – removal interface Boundary between the displacement and removal zones (assumed).

dust collector – removal zone Zone within the dust collector where the dust particles accumulate and may be considered removed from the carrier medium.

dust collector – scrubber Transfers the dust particles in the carrier medium from the fume to a liquid; ie a washing process.

dust-holding capacity The weight of dust which an air filter can retain at is rating as its resistance increases between the 'filter clean' and the 'filter dirty' conditions.

dust – sander See *Sander dust*.

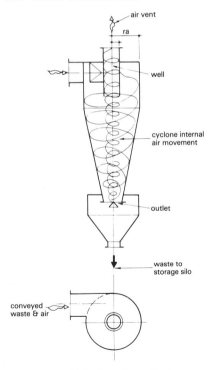

air vent

ra

well

cyclone internal air movement

outlet

waste to storage silo

conveyed waste & air

Figure 38 Cyclone dust collector

dusts – explosive Inflammable dusts, usually in the form of clouds of dust, will ignite and explode under certain conditions of concentration and temperature. Dusts of particular interest to the environmental engineer are wood waste, sander dust, coal dust, cellulose acetate.

Explosions can incur in clouds (accumulation) of dust. The concentration below which explosions are unlikely to occur is of the order of 0.2 to $0.4kg/m^3$. Minimum temperatures below which ignition is unlikely are 600°C (1,112°F) for coal dust, 447°C (837°F) for cellulose acetate.

Dutch oven Over-sized combustion chamber located beneath a waste-heat boiler. Can burn hand-fed off-cuts, logs, paper, certain refuse, etc. (See figure 39.)

dwell time That period of time which a suspended fuel particle spends within the combustion chamber.

dx coil Refers to air conditioning; denotes direct expansion coil.

dx system Direct-expansion cooling system. See also *Direct-expansion cooling.*

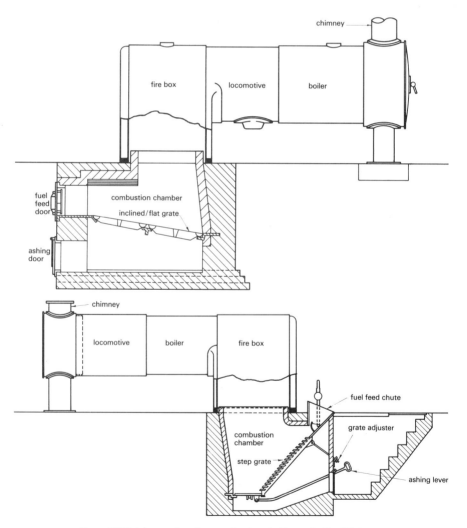

Figure 39 Dutch oven type locomotive type boiler underfired flate grate

earthed circuit A circuit intentionally connected to earth at some point.

earth electrode A conductor which provides a means of connecting the electric network to earth.

economic boiler Self-contained shell-type steam boiler without brickwork settings. (See figure 40.)
A development of the Lancashire boiler with the aim of omitting the brickwork setting and the economiser section, reducing occupied floor space and improving thermal efficiency. It comprises a shell in which the water being heated is stored. Convection heat transfer from the combustion chamber is via two or three passes of fire (smoke) tubes, each between 62mm and 75mm in diameter, which traverse the shell from the front to the back of the boiler. The smokehood through which the gases leave is either at the front or at the back, depending on whether boiler is two-pass or three-pass type. The fuel is fired into the furnace tube, of which a boiler will have one or two depending on size (output).
Manufactured for outputs of up to about 14,000kg of steam per hour at maximum working pressure of 17 to 20 bar.
Occupies about half the space which a Lancashire boiler of same output would cover. Various adaptations have been made to the basic economic boiler type, such as the super economic and the packaged boiler models to obtain higher efficiencies and greater compactness. Used widely for industrial steam raising. Boiler may be of horizontal or vertical pattern. Good water treatment, flue-gas cleaning and blowdown practices are essential to a reasonable life expectancy.

economic thickness Relates to the thermal insulation of pipes and equipment and is that thickness of insulation which, for the given insulation project, is most cost effective. (No thermal insulation provides 100 per cent efficiency; one must therefore balance the overall cost of providing insulation against the saving in heat loss.) Various formulae are published for this exercise.

economiser Equipment principally applied to steam boiler practice; in this, the economiser is in the form of a steel or a copper tube heater battery which is located in the boiler flue system between the boiler flue gas exit and the chimney.
The hot flue-gases which pass through the economiser heat water, which is circulated through the tubes of the battery; in most cases,

the hot water is the boiler feed water, which is preheated in the economiser. Alternatively, the heated water may be put to other uses.
The principle of the economiser may be adapted to other circumstances where hot gases are sent to waste and there is a demand for hot water or hot air.
Application: in steam boiler practice, mainly used with the relatively inefficient boilers (such as Lancashire boilers) which discharge the flue gases at fairly high temperature.

E

eddy current drive Electromagnetic device in which variable speed operation is achieved by varying the slip power dissipated in the coupling by means of adjustment of the excitation to the field system. Efficiency curve decreases linearly with output speed (this is not too serious for pump drives where the load power decreases directly as the cube of the speed).
This drive has a relatively low capital cost and absorbs only a small amount of power (about 1 per cent to 2 per cent) of the drive input, from a small compact solid-state controller.

effective temperature See *equivalent temperature*. Differs from this in that it also takes humidity into account. Requires measurement of the dry-bulb *and* of the wet-bulb temperatures. Widely used in the USA for the evaluation of thermal sensations.

effective-temperature scale Developed in the USA. Stresses importance of humidity in influencing thermal sensation experienced by humans; particularly valid at high temperatures, when the physiological cooling of the person depends mainly on perspiration (eg in the tropics, in hot industries).
Two scales of effective temperatures were constructed: the basic for men stripped to the waist; the normal one for lightly clothed men. Charts are published which show effective temperature plotted against dry-bulb temperature.

efficiency The ratio of

$$\frac{\text{(energy) output}}{\text{(energy) input}}$$

A measure of the effectiveness of any device.

efficiency – volumetric See *Volumetric efficiency*.

electrical corrosion Caused by stray electrical currents from direct-current conductors (eg from electric trains). Protection achieved by use of sacrificial anodes.

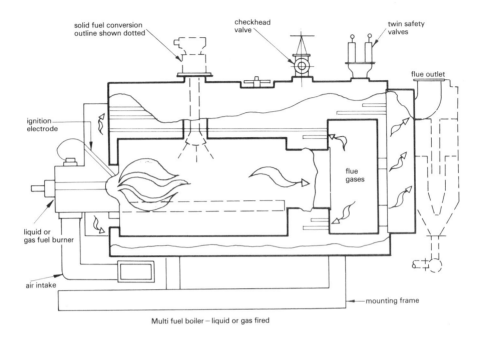

Multi fuel boiler – liquid or gas fired

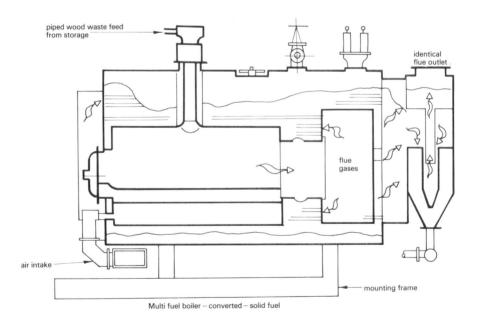

Multi fuel boiler – converted – solid fuel

Figure 40 Multi-fuel boiler

electrical drainage Overcomes electrical corrosion due to stray currents by connecting the pipeline system to a source which has an electromotive force which exactly counter-balances the emf induced by the anticipated stray currents. Such a system will embody automatic-control devices which will adjust the emf immediately the stray currents fluctuate.

electrical scanning Employs infra-red thermographic camera to scan (inspect) the internal condition of electric panels (and similar) and of buried cables with the view to identifying over-heating and faults.

electricity – off-peak See *Off-peak electricity*.

electric motor Machine for converting electrical (input) energy into mechanical (output) energy.

electric motor – alternating current Supplied from an a.c. current distribution. May be single phase or polyphase. The motor speed relates to the frequency of the electric supply.

electric motor – direct current Generally expensive relative to a.c. motor, but used in areas served with d.c. electricity and for special applications where fine speed control and greater flexibility in the motor speed/torque characteristics are required.

electric motor – polyphase Generally three phase. Most motors with outputs of over 1 kw are wound for three-phase supply. A polyphase a.c. supply permits greater output from a given current rating than a single-phase supply on the same voltage.

electric motor – single phase Used generally for motors of up to 1 kw and/or where only a single-phase distribution is available. Most fractional kw motors are wound for single phase.

electric motor – slip-ring Stator and rotor carry polyphase windings. The latter are brought out to slip rings to permit the connection of a separate resistance during start up; having reached the required speed, the rotor windings are short-circuited.
The speed of the motor on any given regulator setting varies according to the imposed load. Starting current is 150 per cent to 300 percent of full load current. To limit the starting current, the motor is usually connected on the lowest setting of the speed regulator. Starting torque at the lowest speed is in the order of 150 per cent to 300 per cent of full load torque.

With a well-maintained slip-ring motor there should be no radio interference.
Available in sizes from 5 to 1,000 kw.
Applications: for larger machines starting against load; in cases where some limitation of the starting current is required.

electric motor – squirrel-cage induction Extensively used on three-phase a.c. supply. Probably the simplest and most widely used of all the types of electric motors.
Consists of a stator wound for a three-phase supply and a rotor of squirrel-cage construction. The rotor winding comprises a series of bars fitted into the rotor slots, and all the bars are connected at each end to a common conducting ring. The bars and the end rings form the 'squirrel cage'.
Has a lagging power factor due to inductive component. Essentially a constant-speed machine, but some speed regulation can be made by voltage reduction.
Available as pole-changing type with two speeds obtained by alteration of the connections to the stator windings.
Application: for high-speed and/or arduous operating conditions. Fairly low-cost motor requiring only little maintenance. Associated with relatively cheap control gear.

electric motor – synchronous or autosynchronous induction The field of such a motor is supplied with a.c. and the rotor with d.c., which is usually provided by small generator (or exciter) driven off the motor shaft. Motor is not self-starting and must be run up to speed by auxiliary windings or by an auxiliary motor. Has a leading power factor which can compensate for the lagging p.f. of induction motors supplied for the same installation. There is radio interference from the exciter; this generally has to be suppressed.
Application: very large motors (up to 5,000 kw) for large steady power loads and where the leading p.f. is desired.

electric motor enclosure Type of protective envelope provided to motor to suit the particular application.

electric motor enclosure – drip-proof (dp) Ventilation openings to the motor are protected to exclude vertically falling water or dirt.
Application: in areas of heavy condensation. Unsuitable for outdoor location.

electric motor enclosure – flameproof (flp) Constructed to withstand any explosion occurring within the motor enclosure and to

prevent the spread of explosion or flame to the surroundings. Gas-tight bearings and provision of sealing glands to incoming cables are essential.
Application: locations where atmosphere may contain inflammable or explosive dust or gas.

electric motor enclosure – pipe or duct ventilated (pv. pvfd, pvid) Has an enclosed case through which cooling air to the motor is continuously supplied via pipe(s) or duct(s). May be self-ventilating by natural air movement, forced-draught actuated or induced-draught actuated air flow.
Application: in locations where risk attends the emission of smoke from the motor windings into the building.

electric motor enclosure – screen protected (sp) Internal motor parts are protected mechanically to obviate accidental contact. Ventilating openings in motor frame and in the end-shields are protected by wire screen or other perforated cover.
Application: relatively dry, clean and non-inflammable areas.

electric motor enclosure – totally enclosed (te) Constructed to ensure that the enclosed air does not connect with the external air, though not necessarily an air-tight arrangement.
Application: in contaminated area.

electric motor enclosure – totally enclosed – fan-cooled (tefc) Motor cooling augmented by fan driven off the motor shaft.

electric motor enclosure – totally enclosed – separately air-cooled (tesac) Motor cooling augmented by separately driven cooling fan.

electric motor enclosure – weatherproof (wf) Construction permits outdoor location of the motor without need for additional weather protection.
Note: associated electric wiring and switchgear must be similarly suitably constructed.
Application: outdoors

electric motor starter – auto-transformer A two-stage method of starting three-phase squirrel-cage induction motors in which a reduced voltage is applied to the stator windings to give a reduced starting current. Consists essentially of a three-phase star-connected auto-transformer, together with a six-pole switch arranged as three double-pole change-over switches. Tappings provided to give 40, 60 or 75 per cent of line voltage at the motor stator on starting.

Provides better starting method than star-delta where the motor must start under load or where slow acceleration must be avoided. Suitable for motor capacities up to 20 kw. (See figure 41.)

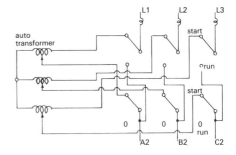

Figure 41 Diagram of autotransformer starter – circuit diagram

electric motor starter – direct-on-line Applies the line current to the electric motor on start up. Cheap and simple. The size of motor which may be arranged for this starting method depends on the permissible peak starting current (starting current likely to be up to 650 per cent of full load current). Varies with different electricity supply authorities and methods; usually limited to motor capacities not exceeding 3 kw. (See figure 42.)

electric motor starter – liquid Liquid electrolyte resistance starter. A recent development which permits smooth continuous acceleration to take place.
Two types available: one utilises electrolyte-electrode resistance; the other employs in addition a vapour stage. Both types are automatically self-variable and offer the advantage of having a fixed electrode, so eliminating a source of wear which is associated with earlier types of liquid starters.
These resistances are inserted in the primary circuit of squirrel-cage electric motors or in the secondary circuit of slip-ring motors. The smooth acceleration prevents mechanical and electrical shocks being transmitted to equipment and to the electrical supply system.
With a liquid starter, the torque increases steadily from the pre-set breakaway value and runs the motor up to speed, maintaining a fairly constant accelerating torque. This promotes a steady rise in motor speed without peaks or torque or current, so extending the life of motors and machinery, control gear and distribution networks.
Various sizes of starter systems are manufactured: up to 800 Kw for squirrel cage and up to

1,600 Kw for slip-ring motors. (See figure 43.)
Application: above 20 Kw considerably
cheaper than auto-transformer starter.

electric motor starter – star-delta Suitable for
the larger motors. In starting position, the
stator windings are connected in star; in
running position, in delta. The stator winding
connections must be taken separately to the
starter. Method reduces the voltage across the
motor winding on starting in star to about 60
per cent of line voltage.
During normal running in the delta method,
each phase winding receives full line voltage.
Star-delta starter is thus effectively a change-
over switch. Suitable for starting motors of
capacity not exceeding 15 kw. (See figure 44.)

electric motor starter – stator-rotor In this,
resistance is introduced into the rotor circuit at
starting and is gradually phased out as the
motor runs up to full speed. Can be adapted to
speed control by running with a varying
number of rotor resistance sections in circuit.
Suitable for large-capacity slip-ring motors (20
kw and over).

electric water conditioner Fitting that incorporates a metal screen to which is applied a
low-voltage electric current. Water flows
through the screen. Claimed that action breaks
up the scale crystals, which are subsequently

discharged with the outflow. Electric consumption minimal wattage. Effectiveness may
depend on water characteristics.
Application: direct hot-water systems, boiler
feed, etc.

electrode boiler Electric boiler used for
generation of steam or hot water for essential
processes in industry; can be used as stand-by
to allow shut down of main fuel fired boilers in
summer months.
Application: off-peak supply of steam or hot
water for essential services by high-efficiency
electric boilers.

electrolytic action Causes or intensifies
corrosion in a system of water pipes brought
about by the presence of dissimilar metals.
Particularly active: copper in close proximity
to zinc (galvanising) or aluminium in a
circulating hot-water system. In water mains,
severe pitting of iron pipes in the vicinity of
lead joints can occur. Fitting together of dissimilar pipe materials should be avoided.

electrolytic corrosion The cause of deterior-
ation or failure of an assembly of dissimilar
metals (most commonly copper and galvanised
mild steel) in contact with each other, accelerated by the presence of a warm/hot fluid; eg
failure of a combined copper and galvanised
steel pipe carrying hot water within a

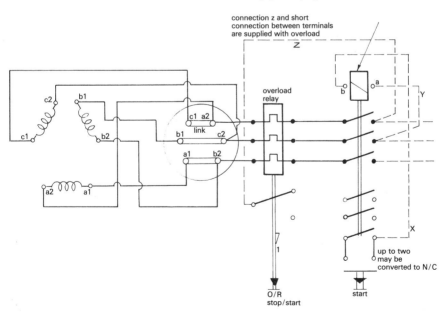

Figure 42 Direct-on-line starter circuit diagram

circulating loop.

electrolytic recovery systems from dilute solutions Can remove soluble pickling and plating process metal remnants in such a way that they can be recycled directly to the plating tank; to be economic, the systems must perform at a cost which is ultimately less than that of the alternative conventional effluent treatment and the value of the metal (eg zinc, silver, copper, nickel, cadmium and lead) which would be lost with the effluent.

electronic heat meter Offers sophisticated measurement and billing of heat consumption to the consumer, including computer billing. Comprises water flow meter section, temperature sensors, calculator and recorder. (See figure 45.)

electrostatic air filter See *Air filter – electrostatic.*

eliminator plate Employed with systems using air washers or spray chambers. Functions to eliminate carried over droplets of water. Construction must suit condition of water being used and the type of contamination being met (eg in industrial and city areas, the eliminators will corrode quickly unless suitably coated or of inert construction, such as glass).

embedded ceiling heating See *Ceiling heating – embedded.*

EMF Abbreviation for 'electromotive force'; causes the flow of an electric current when applied to a circuit.

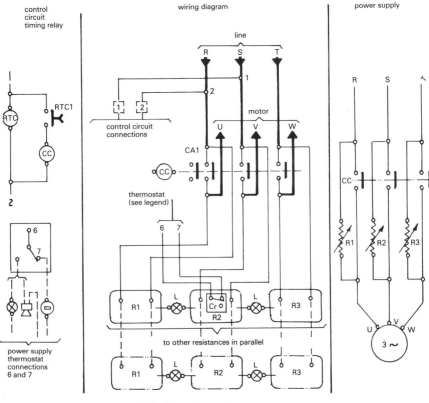

CC – short circuit contactor
CT – changeover thermostat
RTC1 – timed contact on RTC

RTC – timing relay on CC
CAI – auxiliary contact
R1, R2, R3 – thermo variable resistances

Note: in the case of separate coil supply connections 1 & 2 are taken to a terminal block

Figure 43 Liquid electric motor starter-circuit diagrams

emissivity or **reflectivity** Relates to that fraction of heat which is emitted from a hot body by radiation. Depends on the surface finish of that body; it is unity for a perfect black body and less than unity for others.

emittance The ratio of the radiant energy emitted (in the absence of incident radiation) from a given plane surface at a given temperature, to the radiant energy that would be emitted by a perfect black body at the same temperature.

emulsifier Automatically injects the correct amount of water into the oil stream of heavy-fuel-oil burners to create a combustible emulsion. Incorporates pumps, valves and control systems to accurately monitor water flow at all stages of boiler operation. *Applications:* heavy-fuel-oil-burning boilers, kilns and furnaces.

endothermic Reaction or process in which energy is absorbed.

energy – alternative See *Alternative energy.*

energy – auxiliary See *Auxiliary energy.*

energy – geothermal See *Geothermal energy.*

energy – kinetic A moving object possesses kinetic energy (ke) by virtue of its mass (m) and velocity.

$$ke = \frac{mv^2}{2}$$

energy – low-grade See *Low-grade energy.*

energy – potential A body or substance at rest contains potential energy by virtue of its height above a reference level or because of energy stored within it. Unburnt fuel can be considered to have potential energy by virtue of the heat energy locked up within the fuel.

energy – renewable See *Renewable energy.*

energy – total Of a body or substance is the sum of the potential energy and of the kinetic energy it possesses.

energy – wave See *Wave energy.*

energy audit Establishment of energy supply and use balance sheet with primary object of identifying energy-conservation targets.

energy farming Husbanding of field crops and trees for energy and biomass production.

energy management and building control system See *Building energy management systems.*

energy recovery wheel See *Heat wheel.*

energy survey Inspection of an energy-using system with the view of identifying wasteful practices and excessive heat losses. In the UK the Department of Energy sponsors 'one-day' energy surveys which serve to highlight major areas of excessive heat loss and specify in

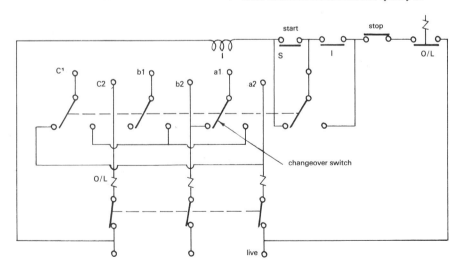

Figure 44 Star-delta starter circuit diagram

outline energy-conservation measures.

enthalpy A property of substances that is a measure of their heat content; it is especially convenient for establishing the quantity of heat necessary for certain processes. The enthalpy of water at 0°C is zero. The enthalpy of steam at various pressure/temperature conditions is listed in the conventional steam tables. Unit of measurement: J/kg.
Enthalpy is a compound function formed from other more simple ones and is defined in integral form by the equation:

$$H = E = pv$$

in which H is the enthalpy (or heat content), E is the energy content and pv is the product of pressure and volume at a particular condition. In general, H should not be confused with the quantity of heat being transferred, although in certain special cases it does become identical. From the above definition, enthalpy is clearly a property; hence changes in its value are determined solely by initial and final states.
In the isothermal expansion of an ideal gas, there will be considerable transfer of heat, but no change in enthalpy.
In the reversible adiabatic expansion of a gas, the heat being transferred is zero, whilst enthalpy increases.

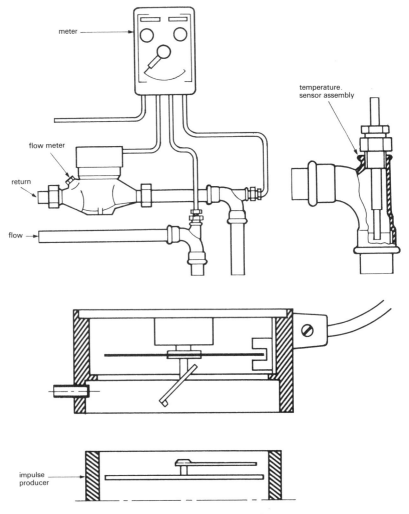

Figure 45 Kamstrup-Metro electronic heat meter

entrainment Conveyance of a fluid by the flow of another fluid flowing at high velocity. Practice commonly used in ventilation and air conditioning, eg an induction system in which the high velocity primary air entrains the main (secondary) air distributed into the conditioned space.

entropy Function first introduced by Clausius in 1851; later developments by Boltzman, Planck, G. N. Lewis and others.
Primarily a mathematical function that offers the simplest means of making quantitative application of the second law of thermo-dynamics.
The environmental engineer will find it convenient to regard entropy as a measure of that portion of the heat energy transferred which is *unavailable* for work.
The unit of entropy is J/°K; the unit of specific entropy is J/kg°K.
For any cycle involving heat and work effects, however complex, but always executed in a *completely reversible* manner, the algebraic sum of all heat effects, divided by the respective absolute temperatures at which the transfers occur, equals 0. This can be expressed mathematically:

$$\frac{Q}{T} = \frac{dQ}{T} = 0$$

where Q is the heat transferred, and T is the absolute temperature.
In a cycle of changes, the total change of the quantity whose infinitesimal value is

$$\frac{dQ}{T} \text{ is } 0,$$

meaning that the quantity has a definite value characteristic of a given state; it is therefore a property. This property, whose derivative is

$$\frac{dQ}{T}$$

is named entropy.
Calculations generally relate to *changes* in entropy.

equal friction method Of sizing ducts, in which the value for the friction loss per unit length of duct is selected and held constant for *all* duct sections of the duct system. Method takes into account the maximum permissible velocities in the various components and sections of the system to obviate excessive noise and power consumption.

equinox The moment at which the sun appar-ently crosses the celestial equator – the point of intersection of the ecliptic and the celestial equator when the declination is 0.

equivalent length Expresses the loss of head (frictional resistance) due to the fittings in a pipeline or duct in terms of the equivalent straight length of pipe or duct; simplifies sizing calculations, as the final computation of resistance losses will all be in terms of straight lengths.

equivalent temperature That temperature of a uniform enclosure in which, in still air, a clothed human body would lose heat at the same rate as in a room or position under observation.

equivalent-temperature scale Used to evaluate the effects of heating appliances in any climate where such appliances are needed. Takes into account dry-bulb temperature, rate of air movement, radiation from or to surroundings. The latter may be either positive (as from exposure to a heater) or negative (as from exposure to a large cold surface).
Is calculated from observations using the formula:
$t_e = 0.522\, t_a + 0.478\, t_w - 0.01474V^{1/2}$
$(100 - t_a)$ where
t_e is the equivalent temperature (°F)
t_a is the dry-bulb air temperature (°F)
t_w is the mean radiant temperature (°F)
V is the air velocity (ft/min)

Escher recuperators Radiation and hollow fin convection recuperators for heat recovery from high-temperature waste gases available with support structure and protection controls. Also available as part of complete combustion system.
Application: to supply pre-heated air to combustion burners on steel reheat furnaces, glass melting tanks, aluminium melting furnaces, kilns, etc.
Energy-saving potential: dependent on the amount of preheat given to the combustion air, but the energy-saving potential is in the region of 20 per cent to 40 per cent giving an average payback of 18 months.

evaporation heat meter Available in two basic types:

to measure the amount of space heating consumed at a radiator;
to measure the amount of domestic hot water consumed.

Meter is based on the rate of evaporation of the meter liquid, being proportional to the heat usage. Each meter must be matched closely to the radiator being metered.

Meters are cheap and, if correctly installed, serviced and read, can provide a reasonable guide to consumption of heat. Meticulous identification of each meter is essential.

Best suited to the *proportioning* of heat usage in a building or a group of buildings. (See figure 46.)

evaporative cooling Technique which relies on its cooling effect on the evaporation of water into a body of air; effect achieved by abstraction of the latent heat of the evaporating water from the air.

evaporator Generally associated with refrigeration and cooling equipment. A heat exchanger in which a liquid is evaporated to

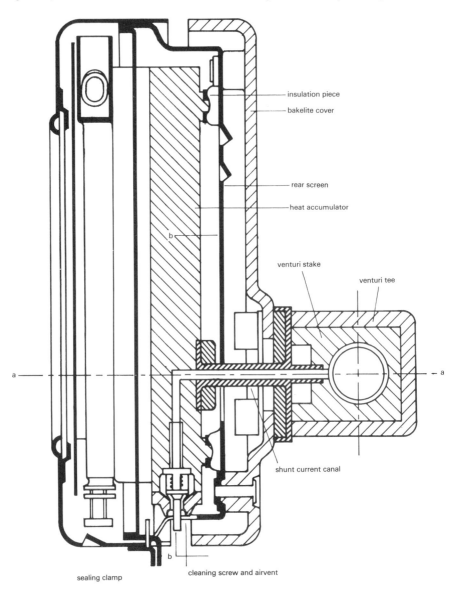

insulation piece
bakelite cover
rear screen
heat accumulator
b
venturi stake
venturi tee
a
a
shunt current canal
b
sealing clamp
cleaning screw and airvent

Figure 46 Clorius evaporation meter

achieve the cooling effect by the process of abstraction of the latent heat of vaporisation from the medium to be cooled (eg air, water, brine, earth).

evaporator unit Basic unit for connection on site to a condensing unit. Incorporates air-cooling coil, expansion valve, fan and motor assembly, air filter(s) and supply and extract grilles. May incorporate flanged connections for attachment to ducting. Drip-tray and drain are required to collect and discharge the water of condensation run off the coil. Evaporator units which operate around or below freezing point require means of de-frosting the coil; usually electric resistance heater(s).

excess air Is required over and above the theoretical air requirement for combustion to achieve, *in practice,* thorough mixing of the air with the fuel. The optimum proportion of excess air varies with the type of appliance, nature and calorific value of the fuel, bulk density, moisture and ash content. It varies from, say, 30 per cent for solid fuel bed type furnaces to 100 per cent for cyclone combusters.

excess air for combustion Quantity of air over and above the theoretical combustion air supply for a particular fuel to ensure the thorough mixing of the air with the fuel. Thorough mixing cannot be achieved by supplying only the theoretical minimum air quantity. The quantity of excess air require-ment theoretically varies with variations of calorific value, bulk density, moisture and ash content of the fuel being burnt and with the particular method of incineration. E.g. solid fuel bed type furnaces require some 30 per cent to 50 per cent excess air quantity; cyclone combustion furnaces require as much as 100 per cent. Careful control must be exercised over the exact quantity and distribution of combustion air (including excess air): inadeq-uate air supply causes smoke emission; exces-sive air supply reduces combustion efficiency.

expanded polystyrene insulation Supplied in various grades and densities in standard sizes 12mm to 500mm thick. Also supplied in linear runs.

expansion – cubical See *Cubical – expansion.*

expansion hose Typically rolled from seamless brass hose with three depths of convolutions, sheathed or unsheathed. May have flanged or screwed ends; soldered or mechanical solder-less connections. (See figure 12.) Suitable for chilled and cold water, low-pressure and medium-pressure hot water, steam and condensate, hot-water service and oil installations.

expansion joint – sliding See *Sliding expansion joint.*

expansion tank (cistern) Associated with open-heating systems and provides space for expansion of the water being heated. Comprises tank (cistern), ball valve with copper float, overflow connection to outside and make-up water connection. Vent pipe usually terminates above surface. Cover should be provided.
Tank size to suit the water content of system. Float control must be adjusted to permit adequate volume to accommodate the expand-ing water. If volume inadequate or float set too high, part of the expansion volume will over-flow to waste, and hard water will be admitted into the system.
Vent pipe(s) must under no circumstances of operation dip below the water level in the tank. Balance of the system must be such that no water is discharged into the tank through over-pumping.

expansion valve Relates to refrigeration systems. The purpose of the valve is to control the flow of the liquid refrigerant (which reaches it from the high-pressure condensing part of the circuit) into the low-pressure evaporator. The pressure drop is achieved by passage through a variable-flow orifice which may be operating in a two-position or in a modulating mode. The classification of expan-sion valves relates to the method by which they exercise control.

expansion valve – capillary tube restrictor For use in small systems (eg domestic size units); the more costly variable expansion valve orifice is commonly replaced by a long small-diameter tube.
This operates as a fixed-flow (non-modulating) control device and will provide reasonably effective control over a wide range of condi-tions. The mass flow through the tube will relate to the pressure drop and to the degree of sub-cooling on entry into the restrictor.
the length and diameter of the restrictor must be closely matched to the operational condition and requirement. Restrictor lengths of between 1m and 4m are common; required tube dia-meters are small: 0.8mm to 2mm.
In view of the above qualification regarding the

selection of restrictor tubes, these are always factory-fitted and tested before dispatch of the packaged equipment to the user.

expansion valve – for dry-expansion circuit In such a circuit, there is no liquid level by which a float valve could be controlled; also it is essential to eliminate any risk of liquid refrigerant returning to the compressor. Such systems are therefore operated with a degree of superheat to ensure the dryness of the refrigerant gas entering the compressor.
Expansion valves are of the thermostatic type, capable of detecting the superheat of the gas leaving the evaporator. They are composed of a temperature detector and a power element charged with the same refrigerant as in the circulation. On falling superheat, the valve closes; if the load on the evaporator increases, there will be more superheat at the detector position and the valve will open wider.
Thermostatic liquid level control and electronic control methods are also available for use with dry-expansion circuits.

expansion valve – high-pressure float valve Given a single-evaporator flooded arrangement, a high-pressure float valve can be fitted which passes the liquid refrigerant from the condenser directly into the evaporator. The float chamber then operates at the higher condenser pressure. This method is unsuitable for multiple evaporator systems and necessitates close control over the amount of the refrigerant charge, as this must not exceed the working capacity of the evaporator.

expansion valve – low-pressure float valve and switch Used in conjunction with *flooded* evaporators, which require a constant liquid level to

ensure that the tubes remain wetted.
A simple float valve is adequate for this application. It must be located external to the evaporator shell to avoid damage to the mechanism by the agitated boiling within the evaporator. Usual method of fitting is by balance pipe connections to the evaporator shell.
The float valve should be supplemented by a tight-closing solenoid valve to stop the refrigerant flow into the compressor when this is stopped, the float valve acting only as a metering valve. This solenoid valve can also be employed to effect on – off operation in conjunction with the float valve; a second float chamber is required to protect against overfilling in the event of failure of the working float control.

explosion The rapid instantaneous expansion of air caused by a sudden abnormally high release of energy.

exposure Relates to environmental conditions of a site, building, wall, window, etc with respect to sun, wind and general weather conditions. Of particular importance with respect to heat gains and heat losses, exploitation of wind energy and requirements for thermal insulation.

extended finned surface By extending the area of metal surface attached to a tube conveying a fluid, the amount of heat transferred to the liquid is increased. Fins may be rectangular or circular, attached to the tube by a mechanical or soldered thermally effective bond. (See figure 47.)

extothermic Reaction or process in which heat (energy) is given out.

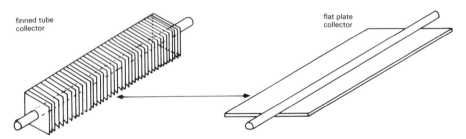

finned tube collector

flat plate collector

Figure 47 Solar collector-finned tube or flat plate

fabric ducting See *Ducting – fabric*.

fabric type air filter See *Air filter – fabric type*.

Fahrenheit Temperature scale widely used in the Anglo-Saxon countries. Is being phased out in the UK in favour of Centigrade scale. Unit is '1 degree Fahrenheit' or 1°F.
See also *Temperature scales*.

fan Conveys air in a controlled and efficient manner against the resistance to air flow offered at the fan outlet; may be direct discharge to atmosphere or movement into an air-handling system.

fan – axial flow See *Axial flow fan*.

fan – centrifugal Consists of an impeller running inside a casing having a spiral-shaped contour; the air enters the impeller in axial direction and is discharged at the periphery, the impeller rotating towards the casing outlet. Pressure developed by fan depends primarily on the angle of the fan blades with respect to the direction of rotation of the impeller.

fan – centrifugal (backward blade) Blade tips incline away from the direction of rotation. Blade angle less than 90°. Non-overloading characteristic.
Application: large-size ventilation systems.

fan – centrifugal (forward curve) Blade tips incline towards the direction of rotation. Blade angle greater than 90°. Develop the highest pressure for given impeller diameter and speed. Usually has between 30 and 60 blades on impeller. Characteristic shows that a small increase in air volume handled results in large increase in the power requirement. Prone to overloading in use and motor requires over-sizing to obviate fan failure under variable load.
Application: ventilation and air conditioning.

fan – centrifugal (radial blade) Blade tips (or the whole blade of a paddle-blade fan) are radial. Blade angle is 90°. Can be particularly stoutly constructed.
Application: conveyance of air which conveys solids in suspension, such as wood chips and sawdust.

fan – characteristic Performance curve of particular fan: inlet air volume against fan static pressure.

fan – double-inlet Essentially two impellers in one compact casing of a forward blade centrif-

ugal fan, giving greatly enhanced output in a small space.

fan – laws 1. *Pressure* developed by fan varies as the *square* of the fan speed.
2. *Volume* of air handled by fan varies *directly* as the fan speed.
3. *Power input* varies as the *cube* of the fan speed.

fan – propellor See *Propellor fan*.

fan – static efficiency The ratio

$$\frac{\text{Static air power}}{\text{Measured fan input power}} \times 100 \text{ per cent}$$

fan – static pressure The difference between the fan total pressure and the fan velocity pressure.

fan – total efficiency the ratio

$$\frac{\text{Total air power}}{\text{Measured fan input power}} \times 100 \text{ per cent}$$

fan – total pressure The difference between the total pressures at the fan outlet and at the fan inlet.

fan – velocity pressure The fan pressure which relates to the average velocity at the fan outlet.

fan-assisted off-peak heater Comprises heat-storage bricks and fan with associated room thermostat. Casing of heater is thermally insulated to high standard. Charges with heat during the off-peak period. Dissipates heat mainly when fan is actuated by room thermostat. Some heat emission (to 10 per cent) from casing.
Application: commercial and domestic space heating.

fan bearing Ball or roller (or sleeve) arrangements can be provided. Current practice favours the ball bearing, as it has been found that roller bearings are more prone to failure, particularly when the fan assembly has been stored for some time before being put into service.
It is good practice to periodically and regularly turn the impeller of a stored centrifugal fan by hand to avoid excessive pressure being exerted on any one part of the bearing, which may cause deformation.

fan-coil air-conditioning system Utilises fan-assisted heat-exchanger cabinets, which are located within the conditioned space. Each unit

F

can be separately controlled by room thermo-stat. Can be mounted on floor or inside suspended ceiling; latter requires supports and access arrangements.

May be of the four-pipe arrangement, which has a separate heating pipe circuit and a separate chilled water circuit. Two-pipe arrangement operates in one of two modes: heating during the winter and cooling during the summer; there is a change-over facility for switching between these modes.

Two-pipe system is more suitable to climates where there is a sharp temperature difference between summer and winter; much less suitable to the British climate. Not suitable for applications where cooling and heating is required. Pipe circuits require thermal insulation; must incorporate vapour barrier or chilled water circuits.

Fan-coil systems operate with separate central heating and water chilling plant.

Application: general air conditioning and for residential use. Requires floor or in-ceiling space for the individual units.

fan – diluted gas boiler flue system Obviates requirements for conventional chimney for all sizes of gas-fired boiler plant. Incorporates a fan fitted into the flue duct off the boiler, arranged to draw air from outside the boiler room and dilute the flue gases sufficiently to permit their safe discharge without the chimney facility. System must incorporate pressure differential or flow switch which will hold out the gas burner until an air flow has been established in the flue duct.

Application: gas-fired installations which cannot readily provide a chimney facility or where the existing chimney would require costly repairs or lining to suit it for gas firing. Chimney facility (where available) generally preferable to fan-diluted system. Details of system must conform to Gas Board requirements regarding capacity and terminal location.

fan-driving arrangement for centrifugal fans Commonly one of the following is provided:

Driving Arrangement 1
1. Belt driven.
2. Impeller supported between bearings fitted on each side of fan casing.
3. Pulley overhung.

Driving Arrangement 2
1. Belt driven.
2. Impeller overhung.
3. Two bearings on one side of casing supported by a pedestal.

4. Pulley overhung.

Driving Arrangement 3
1. Belt driven.
2. Impeller overhung.
3. Two bearings on one side of casing supported by a pedestal.
4. Pulley overhung.
5. Motor mounted on pedestal.

Driving Arrangement 4
1. Direct coupled.
2. Impeller overhung on extended motor shaft. Motor supported by fabricated steel base.

Driving Arrangement 5
1. Direct coupled.
2. Impeller overhung.
3. One bearing, rigid coupling and motor supported by fabricated steel base.

Driving Arrangement 6
1. Direct coupled.
2. Impeller overhung.
3. Two bearings, flexible coupling and motor supported by fabricated steel base.

Double-Inlet. Double-Width.
Driving Arrangement
1. Belt driven.
2. Impeller supported between bearings fitted on each side of fan casing.
3. Pulley overhung.

fan duty Specifies the air quantity handled by the fan (litre/s, m^3/sec or cfm) when operating against the specified external resistance (fan static pressure) (N/m^2m, mm w.g. or in w.g.)

fan lubrication All larger size fan assemblies require regular lubrication with a specified grade of oil. The system designer must ensure that all oiling nipples are easily accessible and that they are clearly indicated on a maintenance schedule. Where easy access cannot be provided to the actual nipples, extended lubricators should be fitted to terminate at good access points.

Close attention to fan lubrication in design and operation is essential for trouble-free fan life.

fanshaft – seal Some manufacturers can provide (against specific request) a shaft seal to limit air leakage around the shaft. Alternative types of seals are made, a fairly common one consisting of a felt ring complete with cover plate which is bolted on to the casing drive-side.

fan – volume control for centrifugal fans May

be by variable-speed motor (can provide volume control to close limits); radial inlet vane damper (has a characteristic which provides a power saving over the conventional multileaf damper in situations where output variations are required); multileaf arrangement (conventional). The two types of dampers may be manually or automatically controlled.

feedback 1. Flow of information from user of equipment to designer, supervisor, etc.
2. Controlling device which maintains set conditions by measurement of the variations in same.

feedstock Relates to production of bio-gas. The organic material from which the gas is generated by organic decay.

fibre board (wood) – chipboard or **particle board** Made from solid fragments of wood which are held together to form a rigid mass by the use of a synthetic resin binder. Most types of chipboard have densities between 400kg/m³ and 800kg/m³.

fibre board – hardboard A hard type of fibre board, of a density between 800kg/m³ and 1200kg/m³. The production of a fibrous pulp is similar to that used for the manufacture of insulation board. A small quantity of wax emulsion is added to the pulp to act as a water-proofing agent, followed by some 2 per cent to 3 per cent of phenolformaldehyde resin. This greatly improves the strength of the final hard-board.
Water-resistant hardboard is made by impreg-nating the board with linseed oil or tung oil.

fibre board – insulation board Insulation board is a fibre board of a density below 400kg/m³. In its manufacture, greenwood timber is first converted into pulp by grinding and mixed with certain other materials such as sugar cane fibres, repulped waste paper, etc, to impart the desired properties to the finished board. This produces a rigid fibre board with an open structure of low density, which is extremely useful for insulation purposes.

fibre board – properties The following relate to various wood fibre products:

K-Values
THERMAL CONDUCTIVITIES

Type of material	Conductivity in W/mK
Standard hardboard	0.14
Tempered hardboard	0.12 – 0.21
Type LM medium board	0.072
Type HM medium board	0.08
Insulating board	0.044 – 0.050
Bitumen impregnated insulating board	0.046 – 0.063

field test Performance check carried out on an operating services installation (eg pressure, volume, velocity, draught tests).

filter – activated carbon See *Activated carbon filter.*

filter – air See *Air filter.*

filter – deodorising See *Deodorising filter.*

filter – rapid gravity See *Rapid gravity filter.*

fines (wood waste) Sander dust, sawdust and similar fine particles.

finned surface See *Extended finned surface.*

fire-brick bonding Method of joining fire-bricks designed to permit expansion movement under load conditions. (See figure 48.)

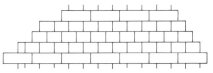

4. Modified English bond

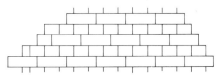

5. Dutch bond

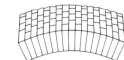

6. Bonded arch

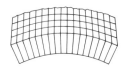

7. Ring arch

Figure 48 Standard brick bonding methods

fire-clay brick Vary widely in properties and in composition. The suitability of this material for any particular application will be determined by the purity of the clay, the absence of fluxes, the aluminia – silica ratio and the grain size. Fire-clay bricks are not generally suitable for high-temperature furnace applications.

fire damper Physical separation introduced into a ducted system and normally held in the open position by a fusible link (low-melting-point material). In the event of fire (or high-temperature condition), the link melts and the damper drops into the separation position, where it prevents the rapid spread of fire along the airway.

first law of thermodynamics The energy added to a system less the energy removed from the system equals the energy change in the system.

flame-failure device Protects combustion equipment (boilers, heaters, process plant) against damage which could arise if an automatic ignition system were to permit continuing fuel input into the combustion space in the event of a satisfactory flame not having been established within a preset time.
Assuming that ignition is *delayed* due to a fault and the fuel continues to pour into the combustion space, a serious explosion would occur if any unduly large amount of the fresh fuel were permitted to accumulate and then lit in an uncontrolled manner.
Such devices may be of the photoelectric (magic eye) type, which views the flame position and will cut off the fuel supply if the flame is not seen within a predetermined time; or it may be a mechanical device which is fitted into the combustion (or burner) space and expands on sensing heat – if it does not *detect* heat within a specified time, it will close the fuel inlet valve.

flame form The form or shape of a flame, controlled by the air pressure and material feed to induce incineration compatible with the shape and size of the given combustion chamber with the object of avoiding flame impingement.

flame stability Maintenance of the correct position of a flame relative to the burner. Essential for well-defined combustion.

flange tables Flanges for pipes, valves and fittings are constructed to suit the particular working pressure of the installation into which the flanges are fitted. The differences in the various flange constructions relate to the thickness and specification of the metal, the number and diameter of the bolt holes and the size of the bolts and nuts used for securing the flanges.
British Standard 10 relates to flanges which are listed as BS Table A, D, E, F, H, J, K, R, S, T, ranging from a maximum steam working pressure of 50 lb/in^2 (Table A) to 2,800 lb/in^2 (Table T).
German DIN flange specifications DIN 2531/2/3/4/5 and DIN 2546/7/8/9/50/51 specify equivalent metric flanges.
Flanges may be constructed of cast iron for pressures up to a working pressure of 700/800 lb/in^2; steel is used for higher pressure applications.

flanking transmission The transmission of sound between two rooms by an indirect path of sound transmission.

flash mixer Chamber in which coagulants are stirred into the raw water with considerable violence induced either hydraulically or mechanically.

flash steam Steam generated from hot water when the pressure is reduced below that corresponding to the saturated steam previously in contact with the water.
For example, low-pressure steam can be obtained from high-temperature condensate discharged from the high-pressure section of a process plant and be used subsequently to service with steam the low-pressure section of that plant.
When condensate is discharged into a vented receiver vessel, flash steam will be generated, and its safe removal must be provided.

flat plate collector See *Solar collector.*

float control See *Control level – float.*

floating control See *Automatic control – floating control.*

floc The fine cloud of spongy particles that forms in water to which a coagulant has been added. The particles are basically hydroxides, commonly of aluminium or iron. They accelerate the settlement of suspended particles by adhering to the particles and neutralising such negative electric charges as may be present.

floor heating – electric Arrangement of electric heating cables embedded within the floor screed. The cables may be so embedded that they are not subsequently accessible or they may form part of a ducted or withdrawable system. Embedded cables can be accommodated within 50mm minimum screed thickness. Floor heating usually operates on the cheaper offpeak tariff. Popularity of this system has markedly declined.

Floor heating – insulated system (plastic pipes)
Arrangement of serpentine plastic pipes developed in Germany. Comprises 13mm diameter 'Multibeton' polyproplene pipes embedded in a specially formulated cement or mortar floor screed. The screed is thermally insulated from the sub-floor and building structure to improve the thermal response of the system. Variable spacing of pipes, which are located on spacing bars, to suit the heat output requirement. 70mm depth of screed required. Water flow temperature 50°C (122°F). Floor coils connected to headers above floor. Suitable for use in conjunction with heat pump.

floor heating – higher flow temperature (metal pipes) Arrangement of floor-embedded serpentine steel pipes of 12mm diameter enclosed inside a proprietary asbestos-cement sheath. Pipe spacing 180mm or 240mm. Maximum flow temperature 71°C (160°F). Floor coils connected to headers above floor. Requires floor screed of 50mm to 75mm thickness.
Advantage: can operate off general-space heating pipe circuit. Heat emission depends on pipe spacing and on floor finish.

floor heating – low-flow temperature (metal pipes) Arrangement of serpentine pipe coils (copper or mild steel) which are embedded within the floor screed above the structural slab. Pipe spacing usually at 120mm, 180mm or 240mm. Maximum water flow temperature 43°C (110°F). Requires floor screed of 50mm to 75mm thickness. Floor coils connected to headers above floor. Heat emission depends on pipe spacing and on floor finishes.

floor panel heating See *Floor heating.*

floor plate Simple or decorative metal or plastic fitting, designed to slip over a pipe which rises through a floor in a visible location and has the object of sealing the gap between the pipe and the floor construction through which it passes.

flow-sheet Diagram which corelates the various activities of a particular process in a simplified form. (See figure 49.)

flow switch Senses, via integral lightweight vanes, gas (air) flow in a ducted system and activates the controlled equipment when flow has been established; conversely, shuts down controlled equipment when air flow is interrupted.
Applications: fan-diluted gas fired boiler systems, critical ventilation systems, etc.

flow – turbulent See *Turbulent flow.*

flue – balanced See *Balanced-flue boiler/heater.*

flue gas recuperation Flue/process exhaust gases are passed through a bank of plain tubes and waste heat extracted by combustion/process/conditioning air which passes over the tubes thus being heated alternatively with secondary surface to heat water or thermal fluid. Manufactured in materials to suit the temperatures and gases being handled.
Applications: industrial furnaces, boiler plant, and process heating.

flue gas tight (damper) isolator Consists of a flanged steel casing housing a hinged blade, which is fitted with a stainless steel sealing strip to give a tight closure. Fully automatic with motor drive for open/closed or modulating control and equipped with explosion relief facility.
Application: prevents cooling effect on boiler of air flow induced by chimney.
Energy-saving potential: reduces pressure/temperature drop in standing boilers and by reducing burner starts gives consequential saving in electricity and wear. Pay back periods of six months to two years are common.

flue stack heat recovery A heat exchanger installed in the flue stack of a boiler or furnace duct. Cold water is pumped through the heat exchanger and is heated by the hot waste flue gases. The flow rate is controlled to maintain after-temperature above the flue gas dew point. (See figure 50.)
Applications: all processes and installations where hot water is required.
Energy-saving potential: 40 per cent to 80 per cent of the exhaust heat can be recovered, the percentage depending upon keeping after-cooling temperature above dew point. Savings of around 17 per cent in fuel-operating costs are not uncommon. (See also *Flue gas recuperation.*)

fluidised bed combustor Shallow bed, atmospheric, two-phase fluidised bed combustors

fitted with patented sparge pipes. In-bed gasif-
ication at 850°C (1,562°F) with complete
secondary combustion above bed producing
gases up to 1,400°C (2,552°F). (See figure 51.)
Applications: for use with shell and water tube
boilers, hot-gas generators for drying
applications and incinerators.
Energy-saving potential: low excess air (10-15
per cent) ensures highly efficient combustion of
conventional fuels and low-grade coal, lignite,
peat, city refuse, industrial waste – solid, liquid
and gaseous.
See also *Fluidised combustion.*

fluidised combustion Fluidised combustion is
a system of combustion in which waste is burnt
continuously in an atmospheric fluidised bed.
The latter is an expanded fuel bed where the
solids are suspended by the drag forces caused
by the gas phase passing at some specific
critical velocity through the voids between the
particles. The solids and gas phases intermix,
acting in the manner of a boiling fluid. (See
figure 52.)

fluidised combustion terminology
AFBC: Atmospheric FBC (applicable for most
industrial boiler installations).
Autothermic balance: The heat release required
to maintain the balance within the bed that
stabilises the bed temperature at optimum level
and equates the heat loss to the submerged
transfer surfaces and the gases exhausting from
the bed.
Bed material: Can be sand, ash from the coal in
use, crushed fire-brick or other similar inert

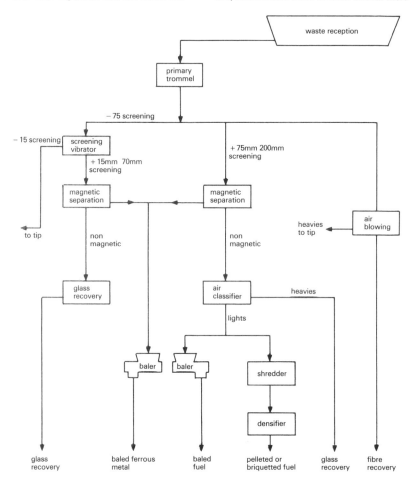

Figure 49 Doncaster plant – process flow diagram

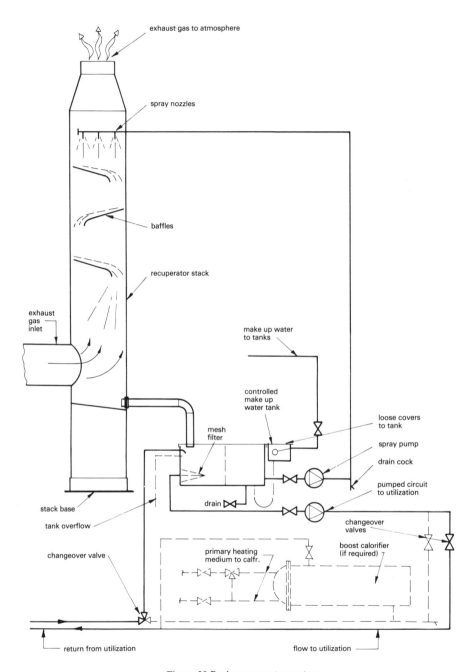

exhaust gas to atmosphere

spray nozzles

baffles

recuperator stack

exhaust gas inlet

make up water to tanks

controlled make up water tank

loose covers to tank

spray pump

drain cock

mesh filter

pumped circuit to utilization

drain

changeover valves

stack base

boost calorifier (if required)

tank overflow

primary heating medium to calfr.

changeover valve

return from utilization

flow to utilization

Figure 50 Basic recuperator system

mineral matter.

Bed refinement: The means to control bed ash content and to maintain bed particle size within prescribed limits. Automatically controlled air-jetting pumps can be employed for bed material cycling in series with screen separators.

Bed slumping: Cutting off the fluidising air supply after post-purge to drop the bed and to cause dynamic and thermal lock.

Bed temperature: Usually limited to 950°C (1,742°F) maximum. At this temperature, clinker formation caused by molten ash in the bed is avoided.

Bed zones: FBCs can consist of a number of bed units or zones, without physical sub-division within the main tank and individually controlled.

Combuster: The complete furnace combustion assembly unit.

FBC: Fluidised bed combuster.

Fluidising gas: The primary (fluidising) air supply, the velocity of which is dependent upon the relationship among the bed surface area, the volume of gases at operating temperature and the bed particle size.

Freeboard: The dynamic active surface zone established above the working bed, usually about 305mm effective depth.

PFBC: Pressurised FBC (for power station boilers, gas turbines, etc where increased combustion intensities are required).

Shallow beds: Bed depths of less than 456mm can be considered as shallow beds. Deeper beds are referred to as deep bed systems.

Support heat: Required to raise the bed temperature from cold and provide a trim heat facility during service. Various methods are used, including over-bed firing and an external form of direct-fired air heater (oil or gas fired) for trim heating purposes.

fluorescent lamp Tubular discharge lamp internally coated with a powder which fluoresces under the action of the electric discharge, producing a shadowless, white or specifically coloured light.

flux A substance mixed with metal to promote fusion, as with silver-solder jointing of copper pipework.

fly ash escape The fine particle escape from suspended matter in the combustion chamber, which has not been carried over to the grit arrestors. Fly ash occurs more usually when the flue gases cool; they become more dense and their velocity decreases, encouraging fall-out of ash and grit. A marginal increase of flue gas velocity may avoid the incidence of fly ash if it is possible to effect the appropriate plant adjustment.

flywheel Heavy wheel associated with rotating machinery (eg steam engine) capable of spinning and of thereby storing energy – attainable energy density up to about 0.2MJ/kg.

flywheel effect Storage of energy obtained by use of flywheel. The flywheel absorbs energy as the machinery slows down and gives out energy when it speeds up.

Used in a wider context, a building of heavy construction is said to possess flywheel effect; it will absorb into its structure solar energy

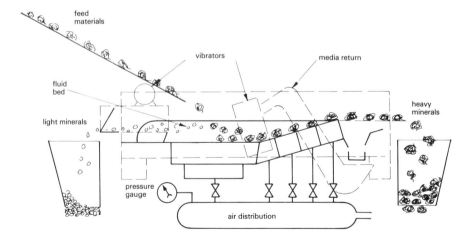

Figure 51 Metals recovery with fluidised bed operation

when this is available and discharge it into the interior as it cools down – thereby retarding the solar heating effect inside the building and the cooling down of the environment when the sun effect has diminished. A similar effect relates to the space heating or cooling of the building by other means.

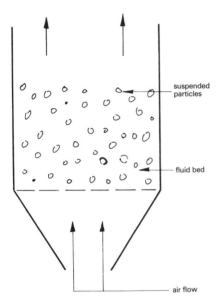

suspended particles

fluid bed

air flow

Figure 52 The basic principle of fluidised bed combustion

foaming Term related to steam boiler plant. Indicates the presence of contaminants in the boiler water, eg concentrations of soap, oil, organic matter, suspended particles, sundry foreign matter. Usually shows in boiler sight glass and confirms unsatisfactory water condition requiring remedial treatment.

foam nozzle Attached to foam pipe and is located in a specified position above the burner equipment.

foam pipes Arrangement of pipes which connect the foam nozzles in a boiler room with a foam terminal box in a suitable accessible external location. In the event of a fire in the vicinity or within the boiler room, the fire service pump fire-quenching foam through the pipes into the boiler room.

forced draught See *Chimney draught – forced.*

forced-draught convector See *Convector – forced-draught.*

fossil fuel Deposit of fuel formed in the earth or under the seabed by the decay of organic matter which has been subjected to great pressure and temperature over a long period of time. The main fossil fuels are coal, oil, gas and the various intermediate stages of fossil fuel formation, such as peat, shale, etc.

four-pipe system Relates to air-conditioning installations. Essentially two separate flow and return circulations, conveying hot water and chilled water respectively to the air-conditioning terminal units.

free air flow Relates to compressed air practice and expresses the compressed air flow at atmospheric pressure (rather than stating the flow at the actual compressed conditions for each application).

freeze-drying Method of preparing dried foods by freezing them and then evaporating the ice in the food directly to steam (ie by sublimation); process takes place at very low pressure.

freon Type of refrigerant used commonly with air-conditioning chillers.

frequency Applied to electrical apparatus; the number of cycles (complete and reverse flow) of an alternating current in each second. Measured in Hertz (Hz).

fretting corrosion Occurs when there is wear and tear between two pieces of steel so that the surfaces tear and the particles oxidise in the presence of air and under the action of heat. Can be a major cause of metal fatigue because it causes cracks to develop in the metal. Likely to affect the bearer units in district heating (and other hot pipe) networks. Is reduced when air is excluded and when the relative humidity is 50 per cent. Can be prevented by lubricating the surfaces with oil, thin plastic or rubber films or by the application of copper, tin or cadmium.

frost damage Water freezes at 0°C (32°F). In so doing, it increases in volume (change of state – liquid to solid), thereby expanding the pipe or container. Frost occurs at temperatures below freezing point, but is not usually revealed until the pipe is 'thawed' or unfrozen.

frost protection Automatic means of adding heat under frost conditions to obviate frost damage.

fuel cell An electrochemical device in which

the chemical energy of reaction between a fuel and an oxidant is converted directly into electrical energy. The basic principle of operation is similar to that of a conventional lead/acid battery; however, whilst the reactants in such a battery require periodic recharging or the battery must be renewed, a fuel cell continues to deliver output as long as the fuel and the oxidant are supplied to it. In a practical application, the reaction between hydrogen (obtained from a hydrocarbon fuel) and oxygen (from the air) are employed.

The fuel cell comprises two electrochemically conducting electrodes separated by an electrolyte (eg potassium hydroxide in water). The fuel (eg hydrogen) is fed to the outside of one electrode, and air or oxygen is supplied to the outside of the other terminal. When the electrodes are interconnected, an electric current flows and fuel and oxidant are then consumed. The electrodes are not consumed during operation, and the fuel cell will continue to supply electric power as long as fuel and oxidant are fed to the electrodes and the products of the reaction are removed. The cell system is refuelled simply by refilling the fuel store.

Theoretical thermal efficiencies of fuel cells can exceed 95 per cent, but in practice these are more likely to be in the range of 30 per cent to 60 per cent; the efficiency loss is attributable to the non-reversibility of the electrode processes and to the power requirement of auxiliaries and control systems. (See figure 53.)

Fuel cells maintain their operating efficiency over a wide range of load variation, and the fuel consumption during periods of idling is far less than with conventional generating equip-ment. Thus, the fuel cell has a superior low-load performance, which greatly increases its versatility of use.

The moving parts of a fuel-cell system are essentially rotary devices of low power which can be manufactured to a standard of reliability and quietness which is suitable for domestic applications. The power-generating part of the fuel cell has no moving parts. Maintenance requirement is much less than for other power-generating plants.

The fuel supply to the fuel cell must be free (or freed) of sulphur products, which would, if present, contaminate the catalysts. Since the combustion process is continuous and carefully controlled, the exhaust products are cleaner and contain less carbon monoxide than those from conventional internal combustion engines.

To achieve a meaningful power output, it is necessary to connect a number of fuel cells in series to form modules, or basic units for constructing higher power systems. A typical module is likely to comprise 40 fuel cells connected in series and is capable of generating in excess of 7kw at 28 volts d.c. Such a module will weigh about 40kg and occupy about 37 litres.

Application: it has been reported that a total of 45 fuel cell systems will be installed at commercial and residential sites in the US by the middle of 1983 by United Technologies at South Windsor, Connecticut. The $32.1 million contract to fabricate and support the testing of the power systems will be funded jointly by the US Department of Energy and the Gas Research Institute. Each fuel cell system will be capable of generating 40 kw.

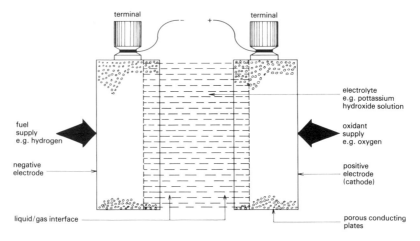

Figure 53 Principle of the fuel cell

Delivery and installation of the systems is expected to begin in mid-1983.

Each fuel cell will produce heat for a commercial or residential building. It is estimated that the combined output of heat and electricity will lead to the use of more than 80 per cent of the fuel's total energy content (compared with the extraction of just over 30 per cent of a fuel's energy by conventional electrical generators).

The fuel cells will use natural gas, although future systems operating in the late 1980s and the 1990s could derive their energy from coal-base synthetic natural gas or hydrogen.

Each experimental 40 kw system will be housed in a cabinet approximately 2.7m long, 1.5m wide and 2m high and weigh about 3,600kg. Currently, 25 to 30 US utilities are investigating possible locations and will submit recommendations during 1982. Selected sites are likely to include apartment buildings, nursing homes, warehouses, stores, restaurants and recreation facilities.

The fuel cells will be delivered in two stages. The first will involve 20 units and the second, 25 units.

In separating the delivery, improvements can be made in later versions on experience gained in the manufacturing, installation and tests of the earlier fuel cells.

fuel efficiency monitor Fully automatic portable flue gas analyser which simultaneously measures temperature and oxygen content to determine the combustion efficiency of a boiler or furnace.

fuel gas Combustible gas capable of being used as an industrial fuel, though it may have a calorific value below natural gas.

fuel oil additives A range of liquid chemicals suitable for all fuel oils from 35 second upwards and for all types of oil-fired plant. They can be added to the storage tank or injected into the oil lines.
Applications: the additives are designed to improve oil burn-out and to keep boilers clean and efficient. Specific additives are available for sludging, sooting, acid smutting, high- and low-temperature corrosion etc.
Energy-saving potential: improvements in combustion efficiency average 3 per cent to 4 per cent and also bring reductions in maintenance costs.

fuel – waste-derived See *Waste-derived fuel.*

fume cupboard Item of laboratory or manu-

facturing furniture concerned with fume-generating process; incorporates suitable safety-engineered enclosure and extraction fan system.

fume incinerator with heat recovery Uses an air-to-heat exchanger to preheat burner gases or for the heating of air, steam and thermal transfer fluids by using the combusted gases created after incineration.
Application: for the incineration of a wide range of combustible organic fumes and particulates, including organic solvents, phenols, aldehydes, oil mists, sulfides, thinners, mercaptans, rendering and sewage odours, aromatic hydrocarbons.
Energy-saving potential: dependent on the use of the heat recovered but can reduce fuel costs to the incinerator by a minimum of 45 per cent.

fundamental frequency The fundamental frequency of an oscillating system is the lowest natural frequency. The normal mode of vibration associated with this frequency is known as the fundamental mode.

furnace – cyclone See *Cyclone furnace.*

furnace rating Design maximum heat input to a furnace: kJ/s m³ of furnace volume.

fusible link Junction between two metal rods or cables formed with a metal of low melting point (usually 68°C (155°F)).
Arrangement is used in locations where there is a risk of fire; when the link melts, the junction parts and actuates a safety measure, eg permits a fire damper to fall into position to separate sections of an air duct or allows a weight-operated valve to close an oil line. After operation, the link has to be replaced. Some types of protective links can be reset; should be approved by the fire authority.

fusible-link valve Shut-off valve fitted into the oil-supply pipe between oil tank and oil burners (preferably external to the boiler room); operates in conjunction with system of fusible links fitted one above each oil burner and a wire linkage. When the fusible links melt, the linkage slackens and the tension on the valve is relaxed causing it to close.
Valve may be weight-operated or spring-loaded.

fusible plug Relates to boilers. Fusible plugs are generally placed at the lowest permissible water level, subject to the direct radiation of the fire or within the direct path of the

fusible plug

combustion gases, as close as possible to the primary combustion chamber.

A choice of fireside steam actuated or water-side plug depends upon the plate or tube in which it is to be inserted.

The fusible plug has a casing filled with tin alloy arranged to melt at about 230°C (445°F) or other suitable approved metal. The fusible metal melts and sprays steam or water into the combustion chamber to put out the fire in the event of excessive temperature at the plug. Usually an insurance requirement.

ganister A variety of fine-grained sandstone containing a small amount of clay matter. True Sheffield ganister contains 97 per cent to 98 per cent silica. The term is also applied to other silica – clay mixture containing 70 per cent and more of silica.

gas – inert See *Inert gas.*

gas – natural See *Natural gas.*

gas – perfect See *Perfect gas.*

gas boiler – atmospheric See *Atmospheric gas boiler.*

gas detection system Equipment for location in boiler rooms which house gas-fired boilers. A probe detects the leaking gas and stops further gas supply to the boilers by closing a solenoid gas valve. A safety device.

gas diverter Functions to prevent down-draughts in a gas-fired appliance from interfering with the efficient combustion of gas. Essential features in the design of a diverter are prevention of adverse action on the gas flame if down-draught occurs in the flue (or chimney); boiler or furnace; acts as a draught break and prevents wide variations in the amount of air being drawn into the appliance due to variations in wind and weather conditions. Figure 54 shows construction and function of a diverter.

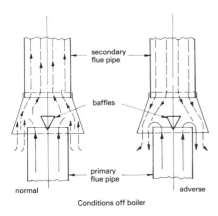

Figure 54 Gas diverter for use with gas-fired appliance

gas-fired infra-red unit heater Direct-fired flueless infra-red heating units using high-temperature stainless steel or ceramic plates as heat-exchanger elements. A variety of control packages available.
Application: for space heating and spot heating, especially in hard to heat areas.
Energy-saving potential: up to 60 per cent of designed heat input is directly used with 30 per cent available as convected heat.

gas-ignition control A fully automatic gas control with electronics used to control the start sequence, eliminating the need for a permanent pilot.
Application: available for use with any gas appliance that requires direct ignition and particularly with appliances in which the combustion air is provided by a fan.

gas main burner Arrangement through which the operating gas supply is fed into the combustion space of a gas-fired appliance. May be a single burner arm or a grid of multiple burners. The gas is ejected through rows of nozzles which are fitted to each burner arm.

gas pilot burner See *Pilot burner.*

gas radiants See *Ceiling heating – gas radiants.*

gas scrubbers Incorporating water sprays are used in connection with incinerator plants to condition the flue effluent, generally to cool the flue gas sufficiently to permit it being handled by an induced-draught fan and to eliminate fly ash from the flue gas system which would otherwise be a nuisance. Each scrubber system incorporates induced-draught fan, fresh water to cool the effluent and means of removal of contaminated water and deposits.

gas turbine CHP installations Combined heat and power installations incorporating gas turbine primemovers with alternator, gas compressor or pump and exhaust heat recovery. Combined cycle installations again incorporate gas turbine primemovers with driven unit, together with steam generator and condensing or back-pressure steam turbine.
Application: CHP and combined cycle installations in the process industries.
Energy-saving potential: overall efficiencies of installations in the UK have been recorded of 60 per cent to 70 per cent.

gas turbine – closed-cycle See *Closed-cycle gas turbine.*

gas turbine – open-cycle See *Open-cycle gas*

G

turbine.

gauge – air filter See *Air filter gauge.*

geothermal aquifer Underground layer of hot water suitable for exploitation of geothermal heat. In the UK such aquifers can be found in certain locations at about 1,500m to 2,000m below ground and at temperatures of 70°C to 100°C (86°F to 212°F).

geothermal borehole Well drilled into the earth's crust to tap reservoir of hot water for subsequent use at the surface. Developing techniques. Existing geothermal working installations in Iceland, New Zealand and France. Present test developments in the UK.

geothermal energy Extractable heat from beneath the surface of the earth. May be in the form of steam (in volcanic regions) or of hot water. Tapped by means of bore holes or extracted from gushing geysers. Exploited in conjunction with district heating and electric power generation.

geothermal fluid Hot water or steam obtained from subterranean strata, by natural geyser flow or from geothermal boreholes. Quality and temperature of the fluid is variable, depending on locality. May have a high mineral content, requiring costly treatment before use in pipeline systems.
Recorded temperatures of fluid up to 200°C (392°F).
Applications: At higher temperatures for steam generation; at lowest temperatures for fish farming, greenhouse heating and similar uses. Heat supply to district-heating systems serving areas close to the geothermal stations.

geothermal gradient Relates to geothermal energy recovery. Defined as the rate of increase of temperature with depth below ground. In a tectonically stable area (such as the UK), the gradient is about 30°C (54°F) per km.
This parameter determines the depth of borehole necessary to achieve useful temperatures for a geothermal scheme and dictates the major cost factor involved in the implementation of geothermal-energy schemes.

greenhouse effect The air temperature under a glass or transparent cover increases when subjected to heat radiation. This 'effect' is caused by the absorption of radiation by the surfaces under this transparent cover and by the ready absorption of radiation in the long-wave length, or infra-red, band being unable to re-radiate through the transparent cover.

grid High-voltage national electricity transmission system operated by the CEGB (in the UK).

grit arrestor A mechanical device which arrests and separates grit and dirt from the flue gases. Usually in the form of a high-efficiency cyclone, operating on the principle of the centrifuge.

grog Previously burnt fire-clay or bricks which are usually ground and incorporated in the clay batch prior to the actual moulding of the fire-bricks. Grog is mainly used to control the drying and firing behaviour of the bricks.

ground water Water which occurs naturally at or below the water table.

group-heating system Serves a number of consumers, who are independently connected to it, usually via a consumer terminal which includes a heat-metering facility. The group system may be connected directly off a boiler plant or off heat exchangers which are on a primary boiler circuit.

glass fibre insulation A bonded insulation of Crown glass fibres, light-weight, strong and free from shot and coarse fibres, easy to handle, cut and install.
Application: thermal insulation in engineering, building and heating equipment up to a temperature limit of 230°C (446°F).
Energy-saving potential: can save up to 90 per cent of heat loss depending on temperature and thickness of insulation used.

glide pipe support Permits easy pipe movement (due to expansion or contraction); eg a roller support.

globe temperature The indication of a thermometer whose bulb is at the centre of a blackened globe – 150mm in diameter; lies between the air temperature and the mean radiant temperature of the surroundings, approaching the air temperature as the air speed over the globe increases. Corelates reasonably well with subjective assessment of warmth.

globe thermometer Mercury-in-glass thermometer arrangement designed for the subjective assessment of warmth. Comprises the thermometer placed within a blackened glass globe of 150mm diameter. Measures globe temperature. (See figure 55.)
See also *Globe temperature.*

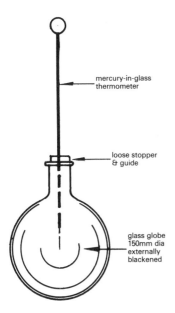

mercury-in-glass thermometer

loose stopper & guide

glass globe 150mm dia externally blackened

Figure 55 Globe thermometer

graphite Used principally in the manufacture of crucibles. Can be added to certain types of fire-clay bricks.

grease air filter See *Air filter – grease filtration type.*

grease interceptor Metal fitting installed intermediate between a source of grease discharge (most commonly a commercial or canteen kitchen) and the public sewer or drain. In its minimum basic form comprises a compartment in which the flow of effluent is baffled and the grease is collected in an accessible container, which is periodically cleaned out manually. More sophisticated grease interceptors are arranged for automatic cleaning.

grease interceptor – actimatic Comprises grease filter and baffle assembly with deep-seal trap, actimatic mixing chamber with dial thermometer and combined cover handle and mixture inlet valve.

To function efficiently, the grease interceptor must be dosed regularly with actimatic powder in solution with water, either on a daily or twice weekly (minimum) basis.

Actimatic powder is a mixture of viable organisms (organisms of propagation) together with bacterial enzymes and food supplements. When the powder is introduced to water, the bacteria immediately multiply and carry on their normal metabolic processes, producing a generous supply of enzymes. Introduced into the mixing chamber of the grease interceptor, the powder permanently converts the fats and grease to water-solution substances and will decompose protein matter and carbohydrates, so that, for all practical purposes, manual cleaning of the grease interceptor is eliminated. Use of the powder also eliminates the unpleasant odours which are commonly associated with fats and grease.

To prime the interceptor at initial operation, a culture must be formed within it. This involves adding a concentrated solution of actimatic powder over a period of two days. This, and subsequent dosing operations, should be made at times when the effluent flow is at a minimum. The initial dosing should be made one to three days after the kitchen commences operation, when the first build-up of grease has occurred. The temperature of the interceptor content at the time of charging with powder should not exceed 43°C (110°F).

gully Outlet through which water or other fluid drains from drained area into the drainage-collecting system.

gully – trapped Gully with a water seal to prevent transmission of smells and vermin.

gully – untrapped Gully without a water seal.

gunmetal Alloy principally of copper and tin. *Application:* where resistance to corrosion or wear required – for the most severe conditions, phosphor bronze is generally preferred.

hammer mill Incorporates a hammer mechanism which reduces the passing material to a specified size. Used in conjunction with waste-recycling plants to prepare the waste for combustion or separation. (See figure 56).

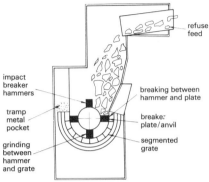

Figure 56 Arrangement of typical hammer mill as used in mechanical sorting plants

hard coal See *Coal – hard*.

hardness Salts having soap-destroying properties (those of calcium and magnesium) are considered in the quantitative evaluation of 'hardness'. Hardness has traditionally been expressed in terms of 'temporary' and 'permanent' components, the former being that proportion of the total precipitated by boiling. Current practice tends towards a more precise evaluation of the water characteristics as 'carbonate' and 'non-carbonate' hardness. Numerically, the carbonate hardness is usually identical with the alkalinity of the water. Expressed as 'parts per 100,000' or 'parts per million (ppm)'.

hardness – permanent See *Permanent hardness*.

hardness – temporary See *Temporary hardness*.

header Denotes a pipe or metal section into which a number of subsidiary pipes are connected.
Applications: flow and return headers connected to a heater or cooler battery; hot-water medium boiler plant (room) flow and return headers, etc. The header is generally of greater cross-section than the connecting branches.

head pressure Relates to refrigeration (vapour) compressors. Is determined by the temperature at which the refrigerant vapour is condensing. An increase in the rate of condenser water circulation and a low condensing temperature will reduce the head pressure, so that, for a given suction pressure, there will be a lower compression ratio, a correspondingly higher volumetric efficiency and a higher overall efficiency.

Health and Safety at Work Act Act promulgated 1974; relates to UK. Intended to provide a broad framework for health and safety measures; supplements previous related legislation. Act established a Health and Safety Executive and Commission and introduced Improvement and Prohibition Notices as alternative enforcement methods to prosecutions. Additionally covers matters relating to the maintenance of the Employment Medical Advisory Services and the promulgation of Building Regulations.
The Act is very widely drawn in respect of the obligations it imposes on everyone concerned with the *working environment*; this tends towards difficulties as regards detailed interpretation. The Act is likely to be eventually tightened up by the issue of related Regulations and Approved Codes of Practice.

heat accumulator See *Accumulator – heat*.

heat balance Provides a balance sheet of the overall heat input and output of a process or system. Particularly useful to establish magnitude and process location of heat losses with the view to minimising them. Can be expressed mathematically or graphically.

heat bridge Formed by a conducting material (usually metal) which bridges two structures – eg the air space of a cavity wall construction – and leads to (unwanted) heat transfer, which increases the heat loss of the structure. It may also act as a conduit for the (unwanted) movement of moisture across the bridge.

heat capacity The quantity of heat which a body takes in when its temperature is raised through 1 degree of temperature. See also *Water equivalent*.

heat dissipator Discharges excess heat into the atmosphere; usually takes the form of a fan and heat exchanger.
Application: in waste-burning applications associated with heat recovery at times when the heat output exceeds the requirement.

heater – baseboard See *Baseboard heating*.

H

heater battery – electrical Assembly of electric heating elements within a flanged casing. Air passes through the casing and is heated by the electric elements; are of the 'black-heat' type and are provided with overheat thermostat to protect against excessive temperature rise. The heating elements for larger batteries are commonly controlled by means of step-controllers to provide heat input modulation.

heater battery – pipe Assembly of parallel pipes connected between two headers of a hot-water or steam system and boxed into a flanged metal casing. The pipes may be plain or finned. In use, the heater battery is connected into a ducted air system (ventilation warm-air heating, air conditioning, etc). The flowing air is heated by passage through the battery. The outlet temperature is commonly controlled by thermostat and motorised valve.

heat exchanger Equipment in which heat exchange is effected between a primary and a secondary fluid; eg air and hot water; steam and water; air and air, etc.
See also *Calorifier, Chilled water battery, Heater battery.*

heat exchanger – plate See *Plate heat exchanger.*

heat exchanger – spiral See *Spiral heat exchanger.*

heat gain calculation Summarises estimates and calculations of all sources of heat gain in a particular environment – eg building fabric, air changes, occupants, electric lights, solar radiation, manufacturing processes, office machinery – usually adopting codified parameters and coefficients.

heating season That portion of the calendar year during which space heating is required and/or provided. In the UK this covers a period of 30 weeks commonly commencing on I October. Other countries have different lengths of heating season.

heating system – sealed See *Sealed heating system.*

heating tapes for pipes Electric surface heating tapes with resistance heating elements providing heat output of 6 to 300W/metre. Tapes for hazardous areas are also available.
Application: protecting pipes from freezing and maintaining process temperatures. Temperatures maintained range from 4°C to 800°C (39°F to 1,472°F).

heat insulation and solar films Consist of two layers of polyester which sandwich a thin layer of aluminium vapour. Normally applied internally to glazing.
Energy-saving potential: it is claimed that winter films save up to 40 per cent of radiant heat. Reflective films turn away up to 78 per cent of solar energy in summer.

heat – latent Heat energy added during the evaporation stage of water to effect the change of state at constant temperature and pressure.

heat loss calculation Summarises estimates and calculations of all sources of heat loss in a particular environment – eg building fabric, air changes, manufacturing process – usually adopting codified parameters and coefficients.

heat meter Allows the direct or remote consumption of actual heat extracted from heating or cooling systems to be read off in kWh over any given period.
Application: heating and cooling systems.
Energy-saving potential: the fitting of a heat meter has been found to be a marked deterrent to wasteful heat use.

heat meter – evaporation See *Evaporation heat meter.*

heat meter – integrating An electronic temperature measuring and flow integrator unit utilised in conjunction with a water meter compensated for hot-water measurement. The integrating counter displays the energy consumption in kWh.
Application: suitable for any application where heat from a central hot-water heating system is to be sold to a user (eg houses) on a district-heating system.

heat meter – mechanical See *Mechanical heat meter.*

heat pipe (chemical) A finned coil fitted with a medium which has a low boiling point. The lower part acts as an evaporator, and the upper part acts as a condenser. It has no moving parts and is of a modular design, hence can be matched readily to air flow and temperature efficiency. (See figure 57.)
Application: designed to recover heat from exhaust air from hospitals, offices, laundry driers, etc.

heat pipe bank Air-to-air units are fundamentally a bank of heat pipes formed into a battery, creating a thermal path between two separate counter-flow air streams. Units are manufactured to specific temperature ranges up to 315°C (599°F).

Application: for use in recovering waste heat that is being exhausted into the atmosphere from drying processes and environmental systems.

Energy-saving potential: recovery efficiency of up to 70 per cent for equal air-flow rates.

heat pipe system Heat pipe recovery systems using heat pipes assembled in arrays to extract heat from hot exhausts from flues and process plant. (See figure 58).

Application: flue recovery systems for boilers, kilns, etc.

heat pump Essentially a refrigerator working in reverse, taking heat energy from a low-grade sink (air, water, sewage, earth) and increasing the level of input energy to effect an increase in temperature; the resultant heat energy can then be applied to heating or cooling uses. (See figure 59.)

Operating in the heating mode, the useful heat is extracted at the condenser. When operating for both cooling and heating, cooling is obtained *in addition* at the evaporator. Suitable change-over valves can be incorporated with the equipment.

The multiple of the energy output related to the energy input at the heat-pump driving mechanism is termed the *coefficient of performance* (COP). It lies between 2.5 and 6, depending on the particular heat-pump arrangement, on the available heat sink and on the operative temperature rise. The heat pump is at its most efficient when arranged for a small temperature rise (eg heating of swimming pools).

Heat pump may operate in one of the following methods:
  air to air
  air to water
  water to air
  water to water
  earth to air
  earth to water

Note: in this case, the term 'water' includes other liquids (eg sewage, industrial wastes). Heat pumps operating in the colder climates must be provided with means of de-frosting or have their operation limited to those times when there is no risk of frost to the equipment. The economics of a particular heat-pump project depend essentially on the available drive energy and on the cost comparison being made. For example, the heat pump will generally prove attractive when using electric drive and competing with electric heating. Substituting, say, bio-gas as the driving medium will greatly improve the viability of a heat pump.

Applications: swimming pool heating, department stores, supermarkets, office buildings, etc.

heat pump dehumidifier Self-contained unit with built-in air-moving system using a refrigeration cycle to remove moisture from the air. Power supply and drainage connections are required.

Applications: process drying, storage.

heat reclaim system Heat-recovery system for new or existing plants with waste heat used for the heating of air, water, steam and thermal transfer fluids.

Applications: for use on all oven and dryer

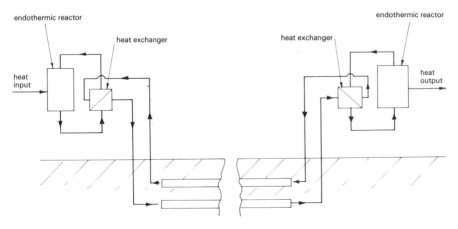

Figure 57 Flow sheet of chemical heat pipe system

exhaust systems for the printing, metal finishing and automotive industries.

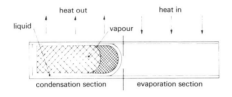

Figure 58 Evaporation heat pipe

heat recovery coil – run around See *Run-around heat recovery coil.*

heat recovery coil – wrap-around See *Run-around heat recovery coil.*

heat recovery condenser pack Draws surplus heat from the compressors of refrigeration plant through the condenser by fans placed at the top of the pack; the warmed air is then fed into an associated ducted heating and ventilating system.
Claimed payback period given as about 2½ years for suitable projects.
Applications: supermarkets, commercial buildings, conference halls, etc.

heat recuperator – plate type Air-to-air heat exchanger available for counter-flow or cross-flow operation, housed in galvanised steel or stainless steel casing with removable covers for inspection and cleaning of exchanger plates.
Applications: environmental ventilation and air-conditioning systems using high fresh air volumes such as theatres, hospitals and swimming pools, etc; industrial hot-air drying processes such as paper, printing, textiles, etc.
Energy-saving potential: the counter-flow type can enable recovery of up to 90 per cent of the thermal energy in the exhaust air, transferring it to the supply air. The cross-flow type is capable of recovering up to 75 per cent.

heat release rate Amount of heat released inside a combustion chamber: $kJ/h\,m^3$

heat – sensible Heat energy which increases the temperature of a fluid (such as water).

heat service contract An arrangement whereby the heat to a building (complex) is provided by a contractor who is bound to maintain an effective heat service under certain conditions based on short-term or long-term obligations and may include the maintenance of the heat-generating (and associated) equipment on a

replacement basis. Such contract usually includes minimum thermal efficiency clauses and some basis for compensation for inflation.

heat-shrink Method used to waterproof the external joints of preinsulated heating pipes, such as employed with district-heating networks. A plastic oversized sleeve is drawn over the insulation at the joint and secured to the joint by the application of heat (by a blow-torch), which shrinks the sleeve firmly onto the joint.

heat sink Arrangement of plant into which surplus heat can be dissipated.

heat store Arrangements which permit storage of heat for future use; eg hot-water storage cylinder, hot rocks, chemical (change of state) substances.

heat – superheat At any given pressure, there is a specific corresponding temperature of saturated steam. Superheat is added at constant steam pressure and raises the steam temperature.

heat – total Relates to the heat content of steam or other vapour/gas; eg applied to steam of a particular pressure and temperature condition, it includes the sensible (heat in the water), latent (heat of steam formation) and the superheat (if any).

heat transfer fluid Specially formulated fluid with specific improved heat-transfer properties; eg tetrasilicates employed for high-temperature heat transfer.

heat wheel (energy recovery wheel) A rotary air-to-air energy exchanger usually installed between the exhaust and air-input systems of a heating, ventilating or air-conditioning installation with the object of recovering up to 90 per cent of the total enthalpy (energy) from the air exhaust stream before this is discharged to atmosphere by transferring this recovered energy to the air-input stream.
Will recover sensible and latent heat.
A typical heat wheel will incorporate an energy-exchanger matrix with a permanent light-weight inert transfer medium. Operates with the moisture content of the air remaining in the vapour state, thereby keeping the matrix dry and free from the risk of bacteria and algae growth.
Insoluble airborne odours are not transferred between the air streams due to the provision of a purge sector in the wheel.
The energy emission of the heat wheel can be

continuously matched to the requirements of the connected system by the use of proportioning controls which vary the speed to achieve the balance. The maximum rotational speed of the wheel is 10 rpm. This low speed minimises the power demand and provides a long life expectancy.
Application: wherever there are large quantities of exhaust air available in proximity of incoming unheated air.
Disadvantage: space requirements and layout of the associated ducting.
See also *Ceramic heat wheel.*

height factor Usually applied as an added allowance to heat loss computations for tall spaces.

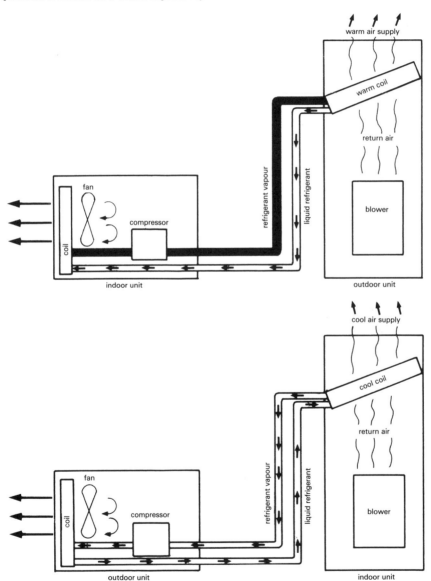

Figure 59 Illustration of heat pump operation

hertz Unit of frequency of alternating electric current – equals one cycle per second.

high alumina Bricks are used for applications which involve high furnace temperatures. The alumina content may vary from 50 per cent to 99 per cent; the latter behaves like a pure compound. High alumina bricks undergo slow plastic deformation under conditions of pressure at high temperature.

high frequency heating High frequency equipment may be divided into three groups, each having its own special field of application:

1. all rotating machines which generate electric supplies at frequencies of 5,000 to 30,000 cycles per second with corresponding power outputs ranging up to several thousand kW, at the lower range of frequencies.
Application: for forging, surface hardening, brazing and soldering.

2. spark-gap oscillators generating at frequencies of 30,000 to 400,000 cycles per second with a maximum power output of 35 kW.
Application: surface hardening of small components.

3. all-electronic valve generators providing supplies of frequencies of 100 kilocycles per second to 200 megacycles per second (the kilocycle and magacycle equal 1,000 and 1 million cycles, respectively). When used for induction heating, generators operating at 100 kilocycles per second to several megacycles per second are employed, with a power output from a few hundred watts to several hundred kilowatts. Most induction heating requirements can be met by generators having an operating range

from 200 to 500 kilocycles per second. Dielectric heating requires the higher frequencies of 1 to 200 megacycles per second. (See figure 60.)

high-pressure/medium-pressure hot water system Functions at water temperatures above the boiling point by the imposition of an artificial pressure head on the system. Alternative to steam system, without complications of steam trapping and condense recovery. Suitable for heat supply to heat exchangers, radiant heaters, process equipment. May employ pressurisation unit or raised static head vessel.
Application: factory space heating, process heat, etc.

high-temperature sealing compound In either ceramic or asbestos form with or without a hardening agent. Suitable for use on its own or with a gasket.
Applications: to effect an efficient air/gas tight seal for flanges, joints, etc, which are subject to heat distortion over time eg boilers, furnaces, ovens, smelters, etc.
Energy-saving potential: minimising gas and air leakages in high-temperature situations; assists in achieving plant design performances.

high voltage Exceeding 1,000 volts on ac system; 1,500 volts on dc system.

histogram Shows the frequency of the occurence of variable values in various ranges in diagrammatic form. In this, rectangles which are proportional in height to the frequencies of occurence, are plotted against a time scale.

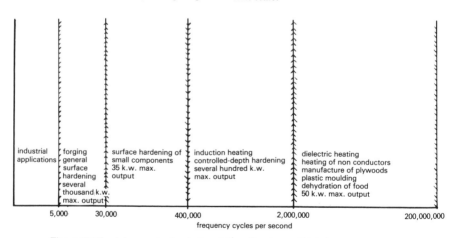

Figure 60 Chart demonstrating industrial applications of high frequency heating

hogged (material) Refers usually to wood chips which have been formed from solid pieces of timber (off-cuts) by passage through a hogger.

hogger Usually associated with wood-waste incineration. Electrically motivated device fitted with chipper knives. Used to reduce size of solid waste to chip size.

holiday A flaw (hole or similar) in the continuity of a protective pipe or duct coating.

holiday detector Electrical instrument for locating a break in the continuity of a protective coating.

hollow plastic floating balls Designed to float and cover any tank containing heated liquids to insulate against heat losses and reduce steam or fumes; also to reduce absorption of oxygen, etc. *Application:* any tank containing chemicals or heated liquids.
Energy-saving potential: a single layer of plastic balls can save up to 75 per cent of heat losses and reduce liquid loss by 87 per cent.

hospital radiator See *Radiator – hospital.*

hot-air disperser Consists basically of a thermostatically controlled fan which automatically operates when the build-up in the roof-void has reached a specific preset temperature. Available in various sizes. One typical range has 15, 20 and 37 kW outputs with effective throw of the air of about 12m.
Application: in high buildings (factories and warehouses) with relatively poor thermal insulation.

hot-water accumulator See *Accumulator – hot-water.*

hot-water boiler See *Boiler – hot-water.*

hot-water meter Permits a constant check on flow in heating circuits.
Application: any hot-water circuit or discharge.

hot-water supply – direct system In this, the tap water first passes through the boiler(s). Such a system is unsuitable for hard-water areas, as scale will then rapidly build up in the boiler water-ways, leading to hot spots and eventual failure.
Towel rails, etc, connected to the direct circuit will suffer air-venting problems.

hot-water supply – indirect system In this, the boiler water circulates through a primary (or indirect) heating coil or heater battery which heats the consumption hot water. Given correct operation, without excessive leakage or emptying of the primary circuit, the indirect circuit should remain relatively free from hard scale. Towel rail or bathroom heaters can be connected to the indirect circulation and will then remain free of air after the initial air has been vented from the system.

hour angle The angular distance of the sun from its position at noon.

$$h = \frac{360}{24} \times T$$

T is the number of hours of the sun time, either side of noon.

housekeeping – good Care of plant in use. Should include preventative programmed maintenance.

humidifier Apparatus which adds moisture to air; eg free-standing portable equipment, duct-located spinning discs assembly, capillary type air washer, spray type air washer, sterile steam electrically operated unit, nozzle type direct steam injector.

humidifier fever An allergy illness caused by exposure to contaminated water from recirculating spray washers. Potential risk offered by spray type using recirculated water and by portable room-located aerosol spray units. The direct-injection humidifiers (eg wet pad drip, spinning disc aerosol, compressed air, sterile water vapour injection) are reportedly unlikely to promote this illness.
Cause appears to be contamination of the water being recirculated due to inadequate equipment cleaning, infrequent change of water after standing idle for a period of days, etc.
Premises at risk: hospitals, centrally air-conditioned offices, printing works, textile mills.

humidity Refers to the presence of moisture in air.

humidity – absolute The weight of water vapour present in a unit volume of moist air, expressed in g/m^3.

humidity – control In ventilation and air-conditioning system is usually achieved by means of a hygroscopic detector activating a moisture-generating device or damper system.

humidity – relative The ratio:

$$\frac{\text{Actual vapour pressure of the air at a given dry-bulb temperature}}{\text{Saturation vapour pressure of the air at the same dry-bulb temperature.}}$$

This ratio is generally expressed as a percentage.
Subjective perceptions regarding the 'feel' of ambient air depend largely on the relative humidity.

humidity measurement Commonly by wet-and dry-bulb whirling hygrometer in conjunction with hygrometric tables. More sophisticated instruments also available.

hunting As applied to services engineering denotes instability about a set condition; eg hunting of controls, hunting of boiler-water level, etc.

hybrids system Relates to collection of solar energy, being a mixture of active and passive solar collection/utilisation systems.

hydrocarbon Any compound of hydrogen and carbon. All fossil fuels are hydrocarbon compounds, usually with added impurities.

hydrofracturing Technique used to improve the permeability of underground strata employed widely in the oil industry and recently applied also to the exploitation of geothermal energy. Consists of pumping water down a borehole until, at pressures of the order of 10 to 14 MPa above hydrostatic, the rock at the base of the borehole splits. Such crack takes the form of a large vertical disc (possibly in a modified form) and can be grown quite controllably by the further application of pressures somewhat lower than were necessary to achieve the original fracture. Such method can attain fracture radii of several hundred metres.

hydrogenation Addition of hydrogen to the chemical structure of a substance. Related to coal, has the object of liquifying the coal and upgrading the extract to petrol/diesel fuels and chemical feed stocks.

hydrolic test Application of a known head of water or water pressure to a vessel or installation conveying a fluid to establish whether there are leaks at the applied pressure.

hydronic system Term for air-conditioning system which uses water as heating and cooling medium (as opposed to all-air system).

hygrometer Instrument which incorporates a dry-bulb and a wet-bulb thermometer in a common frame; supplied with chart that converts the wet-bulb depression to relative humidity. May be suitable for wall mounting, fan aspiration or hand whirling.

hygrometric tables Tabulate the properties of air, usually, for a particular dry-bulb temperature: relative humidity, vapour pressure, dew point, moisture content, total heat, volume, wet-bulb temperature.
Forms basis of psychometric chart.

hygroscopic (material) Readily absorbs or discards moisture from the surrounding atmosphere. Typical materials are tobacco, rayons, textiles, photographic materials, flour.

identibands See *Pipeline identification.*

illumination – efficiency See *Lighting – efficiency.*

illumination – maintenance Regular routine cleaning of luminaires is essential to the optimising of energy conversion to light – particularly important for fluorescent tubes, the output of which is much greater when clean. The efficiency of fluorescent tubes falls off with life (use). Major users of such luminaires therefore tend to adopt a tube replacement programme over a carefully selected period of time to ensure that the overall lighting effectiveness and efficiency is maintained throughout at the highest practicable level.
A number of cleaning firms specialise in undertaking – under contract – such routine programmed cleaning and tube or bulb replacements.

IMechE Abbreviation for Institution of Mechanical Engineers.

immersed crucible furnace A melting and holding furnace in which the metal is heated by a high-velocity burner firing into the crucible immersed in the bath. Heat is transferred through the walls of the crucible into the bath of metal. Temperature is evenly distributed throughout the depth of the bath.
Application: aluminium and zinc alloy die casting and bulk melting.
Energy-saving potential: at maximum melt rate the efficiency is greater than 40 per cent.

immersion heater – conversion Assembly of heat-transfer tubes, so arranged that it can be fitted into or to a liquid-storage vessel for the purposes of making the vessel suitable for operation as an indirect heat exchanger/calorifier.

immersion heater – electrical Assembly of electric heater elements of suitable construction for immersion in water or in other specified liquid; wired for connection to electric supply and, more usually, also to a controlling stem-type immersed thermostat; the latter may have a fixed or adjustable setting.

immersion heater – non-electric See *Calorifier bundle; Submerged combustion.*

immersion heater thermostat Stem-type immersion instrument for controlling water and liquid temperatures.
Applications: hot-water heating installations, industrial premises, oil heaters, etc.

impedance Total *virtual* resistance of an electric circuit or system to the flow of alternating current; arises from the resistance and reactance of the conductor(s).

impeller pump See *Pump impeller.*

incident radiation See *Radiation – incident.*

incineration Disposal of unwanted material in a (purpose-designed) furnace. Reduces the original volume of the burnt material by up to 90 per cent in volume and 60 per cent by weight. Residue: ash. (See figure 61.)

incinerator Purpose-designed furnace to combust waste material. Must conform to clean air legislation and codes. May operate with or without heat recovery. (See figure 62.)
Applications: waste-producing manufacturing industry, municipal authorities, hospitals, abattoirs.

incinerator for gaseous or liquid waste Custom-built fume or liquid incinerator.
Application: any industry which generates fume or liquid waste.

inclined surface A surface or solar-collecting device tilted at an angle to the horizontal plane or to the observer's horizon.

incomplete combustion See *Combustion – incomplete.*

indirect steam heating Applies the heat to a liquid via an immersed heating coil (serpentine or U-shaped) or radiator which is fitted with steam-trapping facility and thermostatic valve control.
Application: for permanent liquid heating, such a system is preferable to direct steam injection; it provides improved steam utilisation and does not dilute the solution. See also *Steam injection.*

induced draught See *Chimney draught – induced.*

inductance Property by virtue of which a change of electric current in a circuit produces a change in flux linkage.

induction air-conditioning system Utilises central heating and chilling plant, which provides heat and chilled water to individual induction units located on the floor or within the suspended ceiling of the conditioned space. The air is supplied into the induction unit at high velocity, discharges through nozzles and

I

101

thereby induces secondary air circulation within the space. The air supply is heated to a controlled programme. Cooling is by chilled water, which is fed through the heat exchanger of the unit. Control by room thermostat and heating programme. High-velocity operation susceptible to noise; careful duct construction and sound attenuation required on ducts and to individual units.

Application: commercial premises. Space required for air ducts to induction units. Relatively high noise level of operation.

induction heating When an alternating electric voltage is applied between the terminals of a conductor, the magnetic field accompanying the current flowing in the conductor is changing with the same frequency as the applied voltage. If a second conductor is placed inside the alternating magnetic field, a voltage corresponding to the rate of change of the magnetic field will be induced in the conductor; if this forms part of a closed circuit, the induced voltage will drive a current through this circuit. In overcoming the resistance of the conductor, work is done which appears as heat. Generating equipment for induction heating commonly consists of two fundamental units. A d.c./a.c. rectifier (complete with the necessary valves) is provided with a tapped transformer, by which the input voltage may be controlled. From the rectifier a high-tension

supply at approximately 7,000 to 9,000 volts a.c. is fed to the oscillator unit, which comprises high-frequency transmitter type valves, transformer, coils and ancillary components. The output from this set is then connected directly to the induction coils, which actually apply the high-frequency power to the material being heated.

The type and form of heating coil used is dependent upon the work to be heated, and the nature and complexity of the process involved. Adoption of a single turn in many cases is found to be convenient, but multiturn coils may be used for mass-production work. The coils need not be cylindrical, but may consist of bends and turns designed to apply the inductive heating effect in certain sections of the component being heated. (See figure 63.)

At the high frequencies employed, the eddy (or induced) currents in the workpiece circulate close to the surface. Thus, given sufficient power in-put and a heating period so short that no thermal conduction can take place inside the body of the article being heated, parts may be heated on the surface only, a facility which is of considerable value where heating through-out may lead to distortion. This phenomenon is termed the 'skin effect', defined briefly as the diminishing of current density from the surface to the interior of a conductor carrying alter-nating current.

The actual depth of heat penetration is a

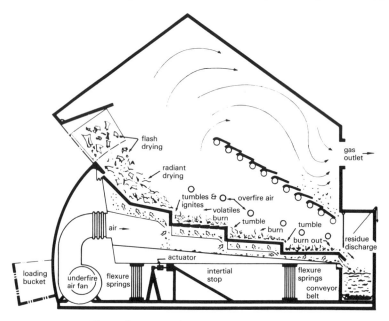

Figure 61 A typical fixed-grate refuse incinerator

function of the applied frequency, but it is also dependent upon the resistance of the material and its permeability. The depth of heat penetration can be precisely controlled by adjustment of the applied frequency.
Application: surface hardening and at the higher range of frequencies for heating articles of irregular shape such as gear wheels where, although the coil is circular, the depth of penetration of heat conforms with the contour of the gear teeth. Similarly, by the application of high-frequency current it is possible to heat an article, such as a cam, by means of a regularly shaped coil and attain uniform heat penetration.

Apart from the advantages of freedom from distortion and avoidance of undesirable internal stresses, high-frequency induction heating enables the heat to be applied locally just where it is required and thereby offers a major advantage over other heating methods. Another field for induction heating is that of soldering and brazing. Where the components are a good fit, the whole operation is performed automatically, the machine incorporating a rotary intermittent movement, so that the parts to be soldered pass successively from the loading station through the heating section, cool and reach the discharge point. When it is desired to solder badly fitting parts, more attention, such as the feeding of the solder by hand, is required.

induction motor See *Electric motor.*

inductive circuit An electric circuit which exhibits self-inductance.

inert gas Does not support combustion and is non reactive. Examples: nitrogen and carbon dioxide.

inertia base Provides vibration control in respect of vibrating machinery. Major advantages claimed: limits amplitude of the vibration motion, gives more stability to system, lowers the centre of mass of the system, provides rigidity among equipment parts, acts as a local acoustic barrier, minimises raction torque effects, provides more even weight distribution, minimises height variation from the the variable loading or reaction forces.

In view of above, tends to extend plant life expectation.

Typical inertia base is a prefabricated steel frame which incorporates isolator brackets and machine fixing bolts; designed to receive poured concrete.

infiltration Random inflow of air into a space via cracks around windows and in the building fabric.

infra-red Invisible long-wavelength radiation (heat).

infra-red camera Employs silicon-coated, nitrogen-cooled lens to photograph thermal images and heat patterns at surface level. Results can be processed as colour or as black and white records.
See also *Thermography.*
Application: energy-conservation surveys.

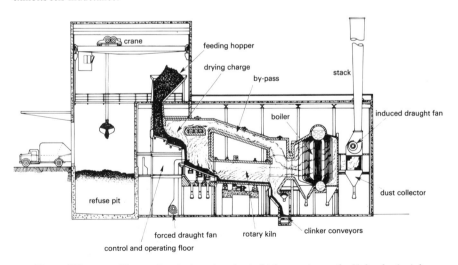

Figure 62 Layout of large urban incinerator plant which operates on the Volund principle

infra-red gas burner

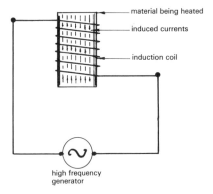
- material being heated
- induced currents
- induction coil
- high frequency generator

Figure 63 Principle of high frequency induction heating

infra-red gas burner Rapid response infra-red gas heater for industrial process heating with face temperature 800°C to 1000°C (1472°F to 1832°F).
Application: preheating of heavy fabrications prior to welding.
Energy-saving potential: energy reductions of 30 per cent can be achieved compared with open flame burners.

injector Injects water into a vessel under pressure condition. Operation relies on the conversion of kinetic energy in the operating steam. Outlet(s) in the form of high velocity jet discharge from a nozzle submerged in the water. Steam requirement: about 1kg of steam to inject 10kg. The steam output will dilute the liquid.

insertion loss The reduction of noise level by the introduction of a noise-control device. Established by the substitution method of test.

insolation Solar energy received at the earth's surface, usually referred to a particular location.

insulating materials – chemical resistance In general, materials which are acidic (pH below 7) should not be used in alkaline environments, and materials which are basic (pH above 7) should not come into contact with acidic materials or such metals as aluminium. Many organic materials have poor solvent resistance. Some insulation materials such as mineral wool induce corrosion in steel, and others may cause a stress corrosion effect with some austenitic stainless steel.

insulating materials – fire resistance The following criteria are demanded for industrial

and architectural insulation materials for safety reasons:
Surfaces must be inorganic so as to be fire resistant. This is of much greater importance than the nature of the body of the insulation. There, at any rate, access of air is restricted so that it does not ignite easily even if it should be heated above its flash point. The only danger which may arise there is that thermal decomposition of the material may take place, producing inflammable vapours which can then penetrate the impervious outer layer to ignite on contact with air.
If inflammable materials are used for insulation, they must be adequately protected by surface insulation using non-inflammable materials. For example, expanded polystyrene, covered by a layer of porous asbestos about 10mm thick, would not catch fire even when a blowlamp is directed upon the asbestos covering for a matter of many hours at a time. It is essential that no toxic vapours are given off by the insulation layer if by some ill-chance it caught fire.

insulating materials – hygroscopy The property of materials to attract water vapour. The condensed water then forms either chemical compounds or associates with it in some way on its surface. In either case, the thermal conductivity of the insulation material rises drastically, and, in addition, there is the danger of corrosion arising upon the underlying metal. Hygroscopy occurs when the water-vapour pressure inside the insulation is lower than that of the surrounding air. If hygroscopic materials are used, they must be fully protected by an impervious layer against water-vapour penetration. In some cases it is best to do the opposite. One should ventilate the insulation layer with dry air to enable water which has been absorbed by hygroscopy to be released again.

insulation – acoustic (sound) Required to limit the perceived sound levels (external to the offending noisy equipment or environment) to a specified noise rating. Achieved by expert use of sound-insulating and deadening materials and special sound attenuating and absorbing baffles, enclosures, duct sections, etc. Insulation may be internal or external to plant and ducts; special requirements relate to the prevention of noise breakout from items of noisy equipment such as fans, hoggers, compressors, etc.
All acoustic insulation materials must be protected against physical damage; where exposed to the weather, they must be weather-proofed to a high degree, as damp insulants

will rapidly deteriorate and fail in use.

insulation – electrical Required primarily for reasons of safety to protect cables and other electrical equipment from contact with other metals (which could cause fire due to short-circuits) and with persons (who might be electrocuted). The insulant is itself protected by an external sheath to obviate mechanical damage, which could expose the bare conductors, and to guard against ingress of damp and condensation.
The insulation system may include a screening cable in communication applications, where the effect of stray currents on the conductors and equipment cannot be tolerated.
Insulation used to be principally by vulcanised rubber (VIR); this is now seldom used, as the rubber tends to harden and turn brittle.
Modern insulating materials are plastics (PVC) and mineral (MICC) systems.

insulation – loose-fill See *Loose-fill insulation*.

insulation – pipe Thermal wrapping or lagging or spray applied to a piped energy-using system to (greatly) reduce loss of energy by convection and radiation.

insulation – rock-fibre See *Rock-fibre insulation*.

insulation (sound) The property of a material or partition to oppose sound transfer through its thickness.

insulation – thermal Required to greatly restrict unwanted heat losses (or gains) from hot or cold surfaces, such as plant and pipe systems. The insulation materials by nature are fragile and must be protected against mechanical and moisture damage as appropriate to the circumstances of the installation. The materials are selected to suit the temperature range of the application. Insulation should incorporate an effective vapour seal when cold surfaces are insulated, eg chilled water, brine, etc.
Common insulating materials are glass fibre, mineral wool, magnesia, asbestos (less common now due to restrictions on its use because of potential health hazard), polyurethane foam, cork and polystyrene (for cold-surface insulation). Various types of protective sheaths are used, such as sheet metal, aluminium, canvas, plastics. In large buildings (such as those within Section 20 in the Greater London area) the local authorities limit insulating materials to those which do not offer any fire hazard.

integrating heat meter See *Heat meter – integrating*.

integrating steam flow meter Employs the pressure differential across an orifice or Venturi meter to determine the flow rate of the steam on the principle that the pressure exerted by a flowing liquid is directly proportional to the square of the velocity of flow (Bernoullis' theorem). The complete metering assembly is arranged to integrate such pressure differential with the steam pressure and to record the actual flow rate (in appropriate units). See also *Sequential metering*.

inverse square law The intensity of radiation (heat or light energy) from a point source of radiant energy decreases with the square of the distance from the point source.

inverse square law – sound The reduction of noise with distance. In terms of decibels, it means a decrease of 6dB for each doubling of distance from a point source when no reflective surfaces are apparent.

inversion Temperature inversion in the atmosphere reverses the normal trend of temperature rise, so that there is no longer the tendancy to form upward currents.

inverter An inverted rectifier for converting direct electric current into alternating current. Operates with batteries and offers an alternative to conventional generator for certain applications, with particular advantages for cumputer systems.

ionizer, ionization Equipment designed to produce a high negative ions concentration within the treated space. This is said to produce environmental benefits and has been monitored to such effect in certain clinical trials.

iris damper Circular damper with 'camera' type damper action, efficient means of balancing duct air volumes due to reducing orifice arrangement, maintains uniform flow over the cross-section of the duct, obviates increase in high-frequency noise when dampering. Adjustment of flow by single extended lever, which by angular rotation through 90° regulates the air volume from full free area to tight shut-off.
Application: at air distribution terminal fittings and inside ducts.

irradiance The radiant energy falling per unit

ITOC plant

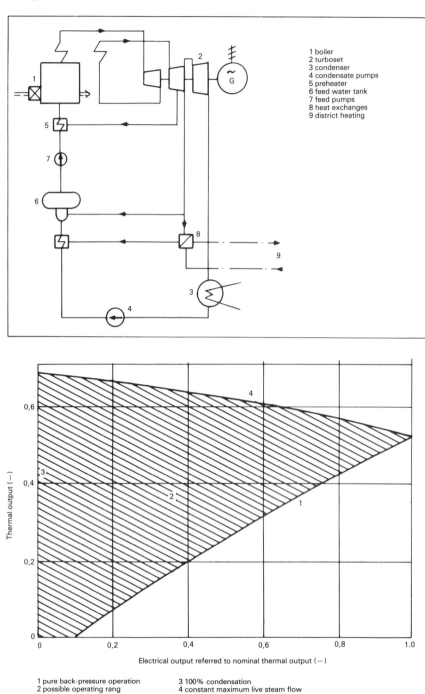

Figure 64 Sulzer fossil fuelled ITOC plant

area on a plane surface per unit time, normally stated in watts/m² or Btu/ft².

isenthalpic Process at constant enthalpy.

isentropic Process at constant entropy.

isobaric Process at constant pressure.

isolation – sound The reduction of airborne sound transfer from one area to another.

isolation – vibration The reduction of vibrational force into a structure.

isolation efficiency The amount of vibration force absorbed by an isolator and thus prevented from entering the suporting structure, expressed as a percentage of the total force applied to the isolator.

isolator (electric) Simple switch for opening or closing a circuit under conditions of no load or of negligible current.

isothermal Refers to a process conducted at constant temperature.

isothermal compression Takes place at constant temperature.

isothermal expansion Takes place at constant temperature.

ITOC stations Abbreviation for 'intermediate take-off condensing' stations. Relates to district heating. Operate electricity-generating turbines with maximum heat bleed when external power demand is low and normally when demand for power is high.
Produce heat in the form of hot water or low-pressure steam a very small cost. By the operation of a single valve, it is possible to adjust the output of the turbine instantaneously from producing power but no heat (at an efficiency almost identical to that obtained with conventional condensing turbines) to producing both electricity and heat up to the maximum design heat output of the plant. The heat output of such plant can be adjusted, second by second, to accord accurately with the heat demand of the system without varying the

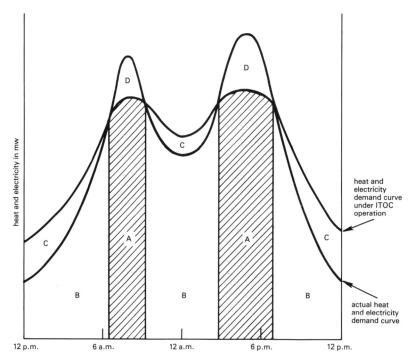

Figure 65 The operation of an ITOC turbine flattens down demand **A** *Turbine under condensing operation only,* **B** *Turbine under partial back pressure operation,* **C** *Extra heat produced is stored against peak demand,* **D** *Heat demand supplied from stored hot water*

fuel through-put of the furnace because such adjustment merely alters the amount of electricity generation sacrificed to produce heat. Such stations obtain between five and six units of heat, on average, for each unit of electricity output sacrificed. (See figure 64.)

ITOC turbine Abbreviation for 'intermediate take-off condensing' turbine. Relates to district heating. In ITOC turbine systems, an automatically controlled valve adjusts the output instantaneously.

Operation of ITOC turbine system effects a flattening out of electricity demand at the station. The turbines are operated at night with maximum back-pressure operation and the surplus heat is stored in accumulators and within the pipelines as sensible heat. During the morning peak demand period for electric power, the bleed valves for steam are closed, and the turbine then operates as a fully conventional condensing turbine, giving its maximum rated output of electricity but no heat. The hot water or low-pressure steam which has been fed into the storage overnight then provides external heat supply and relieves the power demand by lessening the demand for peak current output.

The turbines are operated on maximum-condensing cycle until the power demand lessens around lunchtime. The back-pressure valves are then opened somewhat to boost the water temperature to meet the heating load during the afternoon and evening peak demand periods, when the turbines revert to operation on full-condensing cycle. (See figure 65.) For advantages of ITOC operation, see *ITOC stations.*

kaolin The general term for china-clay rock, derived from Kao-Ling, a high hill in north China from which the Chinese obtained their supplies of clay.

kaolinite A definite mineral form of china-clay of chemical composition $Al_2O_3\ 2H_2O$. When heated above $1,100°C$ ($2,012°F$), it breaks down to mullite, cristobalite and glass.

Kelvin (°K) Name given to the scale of absolute temperature (degree Kelvin). Zero point of the scale is the absolute zero of temperature. The degrees Kelvin within the scale correspond to degrees Centigrade. Conversion from degrees Kelvin to Centigrade is achieved by subtracting the number 273.

kiln Used for the gradual forced drying of timber under controlled conditions. The heating and humidifying medium is usually steam.
Modern timber kilns are highly instrumented to produce a closely specified moisture content in the timber with minimum steam input. Kilns are large users of steam.

kilning Practice of curing timber faster (forced drying) than possible by natural air drying. The process involves the placing of the timber into a drier (kiln) for a specified period (24 hours or more) and drying of the timber under strictly controlled conditions of temperature and humidity.

kinetic energy See *Energy – kinetic*

Kirchoff's Radiation Law The radiating capacity of a given body, represented by the radiation constant E, for given temperature and wavelength, is proportional to the absorbing capacity of the body.

Koanda effect Refers to the presence of a stagnant layer of air which occurs with air movement from ventilation/air-conditioning supply outlets. During cooling, the natural convection currents form a stagnant zone between the stagnant layer and the ceiling; during heating, the stagnant zone is formed between the stagnant layer and the floor. (See figure 66.)

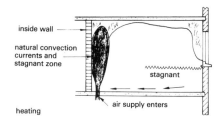

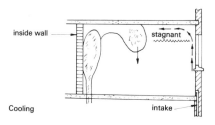

Figure 66 Illustration of Koanda effect

lagging Term used for thermal insulation to pipes and plant.

Lambert's Radiation Law The radiation from a surface in a direction at an angle with the surface varies as the cosine of the angle between the direction of the radiation and the normal to the surface.

La Mont boiler Forced-circulation water-tube boiler suitable for steam generation or high-pressure hot-water service. (See figure 67.) See also *Water-tube boiler*.

Lancashire boiler Brick-set steam boiler with two internal flues shell. Very popular in the coal-burning era due to robustness, simplicity, long life and simple attendance. To achieve worthwhile thermal efficiency, must be fitted with an economiser (see *Economiser*) to

improve the convective heat transfer surfaces. Many Lancashire boilers still in world-wide use. Some have been converted to heat-storage cyclinders or accumulators. Boilers also used for burning wood waste.
Lancashire boilers manufactured in sizes to 3m diameter x 10m long; steaming capacity to 6,000 kg/hr at pressures of up to about 17 bar.

Langley Unit of measurement of solar radiation intensity, particularly used by meteorologists. One Langley equals the radiation intensity of 1 calorie per cm^2.

latent heat See *Heat – latent*.

latent heat of fusion Heat required to change from solid state (say, ice) to liquid state without change of temperature.

L

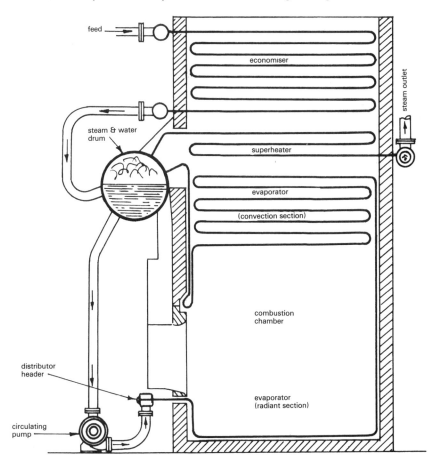

Figure 67 Forced-circulation water-tube boiler – diagrammatic arrangement

latent heat of vapo(u)risation Heat required to change from liquid state to vapour state without change of temperature at constant pressure.

latitude The latitude of a point on the earth's surface is its angular distance from the equator; eg the latitudes of the following cities are:

London   51°30'N
New York   40°41'N
Sydney   33°52'S
Delhi   28°38'N

leak detector For ultrasonic frequencies produced by the passage of a gas through an orifice. Signals picked up are converted into an audible hiss and/or displayed on a meter. It can be used to detect leaks, corona discharge and mechanical wear.
Applications: compressed air and steam leaks; bearing wear; steam trap and hydraulic valve malfunction; high-voltage corona discharge.

leak detector – ultrasonic See *Ultrasonic air and steam leak detector.*

level control – electrodes Actuates controlled equipment by movement of the fluid between the fixed electrode positions.

lighting – efficiency Fluorescent tubes are currently the most popular light sources for offices and factories. They are energy efficient, but the luminaires are more costly than tungsten and mercury ones.
A wide choice of fluorescent tubes is on offer, some more suitable to a particular purpose than others. The conversion of electrical energy to light can differ for alternative luminaires from about 80 lumen/watt to only 30 lumen/watt.
A recent energy-saving development of the conventional fluorescent lamp is a miniature version of which has the fluorescent ballast and starter also miniaturised and fitted into the base of a bulb which may have a bayonet or screw form of attachment. A particular application of this is the 'Philips SL' bulb, which uses phosphors from the Philips Colour 80 series fluorescent lamps, derived from colour-television technology, to offer a high light quality combined with high light output. It is claimed, following tests-in-use, that the SL lamp will reduce the equivalent electricity consumption of equivalent tungsten lighting to one quarter and increase the life of the bulb five times (to 5,000 hours). However, *one note of caution:* before deciding to replace the tungsten bulbs within an existing luminaire with SL bulbs, one must confirm that the different dimension of the latter can be accommodated. (See figure 68.)

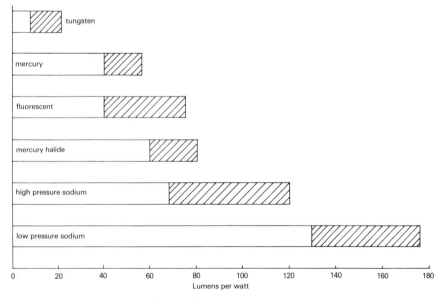

Figure 68 Diagram indicating efficiencies of different popular light sources

Cross-hatched area is the range of efficiencies for the specified light source with different types and sizes of lamps. Note: generally, the larger lamp sizes are more efficient

lighting – maintenance See *Illumination – maintenance.*

lighting control – automatic See *Automatic lighting control.*

lighting control optimiser Photoelectrically operated device with separate on-off controls and including a 7-day, 24-hour time clock for programming. Suitable for direct switching of lighting load up to 20 amps or larger load using separate contactors.
Application: interior lighting installations in factories, offices, public buildings, cleaners, lighting, etc. Flameproof version also available.
Energy-saving potential: energy savings of up to 30 per cent are possible providing payback within two to three years.

lignite See *brown coal.*

lime – soda plant Water softening plant which uses lime and soda ash to make calcium and magnesium salts insoluble so that they form precipitates which can be settled and filtered out of water.

lime – soda water treatment Adds lime and soda ash (if required) to waters with temporary hardness salts. Non-soluble carbonates are then formed which can be settled out and removed from the water system – a precipitation process.

linear expansion coefficient The increase in length which a bar of a material of unit length undergoes when its temperature is raised through 1 degree.

liquid phase boiler Serves heating applications (mainly process ones) which require relatively high temperatures *without* the high pressure which would be associated with steam systems operating at similar temperatures. Boiler operates with manufactured high-temperature heat transfer fluids, such as tetra cresyl silicate. Temperature range without boiling: up to 315°C (600°F).

liquid starter See *Electric motor starter – liquid starter.*

live (electrical) Ascribed to a circuit or conductor of electricity actively connected to a source of emf.

load factor Has maximum value of 1; usually less than 1. The proportion of the design (theoretical) loading of a system (eg district heating, geothermal energy, electric distribution, etc) which applies at a particular time. A high load factor can be crucial to the success of a particular scheme, as the benefits of scale are then maximised.

local apparent time (lat) System of astronomical time in which the sun always crosses the true north-south meridian at noon. This system of time differs from local time according to longitude and time zone. The precise displacement also varies with the time of year.

loft and wall insulation – blown fibre See *Blown fibre loft and wall insulation.*

longitude The angle which is made by the terrestrial meridian through the geographic poles and a point on the earth's surface with a standard meridian (usually at Greenwich, England).

loose-fill insulation Granular loose-fill insulation.
Applications: Roof insulation, back fill for boilers, etc.

loose-fill powder thermal insulation Applies to underground pipe systems. The powder is moisture repelling and has thermal-insulation properties. It is poured into the pipe trench after the pipes have been installed. All support brackets, anchors and the like must be treated with moisture repellent to obviate water channelling into the trench via same. The trench must be free from all rubbish (foreign matter). The thickness of cover over the pipes must conform to manufacturers' instructions (related to their guarantee) preferably with some margin. Manholes, valves and expansion fitting require special care.

loss factor See *Dielectric heating.*

low-grade energy A form of energy which is available in abundance, but cannot easily be harnessed for purposes of energy recovery. Of importance in heat-pump techniques (which upgrade low-grade energy to a usable higher level). Examples include heat in the earth (excluded geothermal), power-station cooling water, rivers, ponds and atmospheric air.

low voltage Exceeding 50 volts ac or 120 volts dc to a maximum of 1,000 volts ac or 1,500 volts dc.

LPG Abbreviation for 'bottled butane and propane gas'. GJ/tonne: 49.3 butane – 50.0 propane. *Latent heat* at 15.6°C (60°F): 372.2

lumen (lm)

kJ/kg butane – 358.2kJ/kg propane.
Volume of gas at 15.6°C (60°F) and 1015.9
mbar: 406 – 403 dm³ butane – 537 – 543 dm³
propane.
LPG is a convenient substitute for piped
natural or manufactured gas in locations where
the latter not available or for temporary works.
More costly than piped gas.

lumen (lm) SI unit of luminous flux that
describes the total light emitted by a source or
received by a surface.

luminaire Widely used term for an electric-
light fitting.

luminaire – efficiency See *Lighting –
efficiency*.

luminaire – maintenance See *Illumination –
maintenance*.

lux (lx) Unit of illumination; the illumination
produced by a light from a source of one international candle falling directly on a surface at a
distance of 1m from the source.

macroclimate Regional climate of an area, usually as reported by the local meteorological station.

magnesite An impure refractory material suitable for high furnace temperatures except when under load and when large amounts of steam are present. It shows poor resistance to spalling, unless it is pure electrically sintered magnesite, crushed, bonded with a magnesia salt and fired. Fused magnesite shows fair resistance to spalling and may be used when subjected to high temperature and the action of basic slags.

magnetic separator Separates ferrous materials by magnetic attraction. (See figures 69 and 69A.)
Application: waste separation, wood waste (similar material) handling to protect the handling plant from damage through the presence of unwanted metal, such as nuts, bolts, etc.

magnetic water conditioning Arrangement of permanent magnets through which the water flows. Claimed that passage through the magnetic field breaks up the crystals of scale; these are subsequently discharged with the outflow of water. No electric connection required. (See figure 70.)
Application: popular in Soviet Union in connection with once-through district-heating systems. Domestic and commercial hot water systems, etc.

magnetohydro-dynamics (heat recovery) In this, liquid metal is circulated through a heat source (eg solar collectors, industrial wastes, sewage) in conjunction with a vapour circuit. The vapour is a highly volatile refrigerant which assists the metal to flow past a magnet at which an electric field is generated and electricity produced. The latent heat of the vapour can also be recovered at the condenser heat exchanger.
This type of system is currently undergoing pilot plant evaluation and appears to possess great potential for heat recovery and low-cost electricity generation. (See figure 71.)

make-up water Related to closed heating systems, make-up water is that amount of water which must be added to make good losses through glands, expansion, pumps, etc. In the open system, such make-up takes place via the feed and expansion cistern; in sealed

M

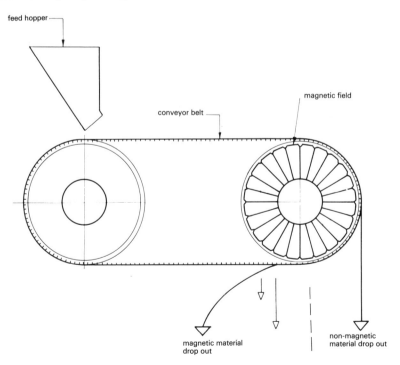

Figure 69 Magnetic separator – conveyor belt arrangement

systems, the water is fed via the expansion vessel circuitry.

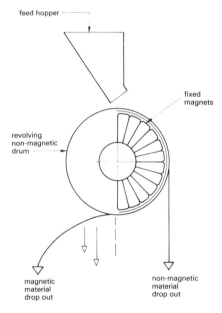

Figure 69A Magnetic separator – magnetic wheel type

manometer Basically a U-shaped tube employed for the measurement of pressure differentials in liquid systems. May incorporate visual or audible alarms to indicate over-run of preset pressure differential.

manual reset Adjusts the set point of the controller to accommodate load changes.

maximum demand (electric) Maximum current or power that has flowed through an electric circuit in a given period. Used in connection with tariffs for electricity consumption, which are based on a specific maximum demand and impose a cost penalty when that demand has been exceeded.

maximum demand (meter/alarm) Monitors the flow of electric current or power through an installation relative to a specified maximum demand. May include a visual or an audible alarm feature to indicate approach of the maximum demand consumption.

maximum demand load controller Automatically measures, predicts and controls electricity usage so that consumption is optimised. It predicts probable total demand in a half-hour integration period. If this predicted demand seems likely to exceed a preset target, an alarm relay is activated, causing either alarms to be energised or loads to be automatically shed. Can take form of a micro-processor-based system, which continuously monitors electric-power consumption and automatically sheds or cycles non-critical loads.
Applications: retail stores, hotels, offices, industrial plant. Commercial and industrial premises on md tariffs.
Energy-saving potential: a saving of 9 per cent to 14 per cent of total electricity bill can report-edly be achieved on industrial applications.

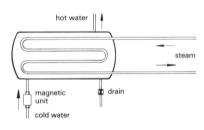

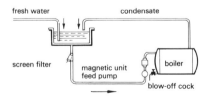

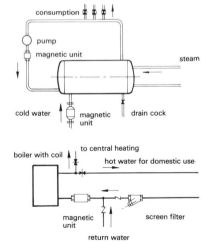

Figure 70 Diagrams showing magnetic water conditioning process

mechanical draught See *Chimney draught – mechanical.*

mechanical heat meter – hot water Measures the quantity of hot water being supplied to the consumer and also the temperature difference between the flow and return flow to and from the consumer. The two parameters are integrated, and the heat consumption, in appropriate heat units, is shown on a readout panel. The degree of sophistication of particular heat meters must relate to the metering needs and to available finance. (See figure 72.)

mechanical seals See *Seals – mechanical.*

metabolism The amount of energy a human being dissipates into the environment. For an adult in temperate surroundings, the average metabolic rate due to all energy losses (radiation, convection, perspiration and exhalation) is in the order of 300 watts.

meter – Btu See *Btu meter.*

meter – condensate See *Condensate meter.*

meter – oil See *Oil meter.*

meter – Venturi See *Venturi meter.*

metering – sequential See *Sequential metering.*

methanol (CH₃OH) A methyl alcohol. Clean liquid fuel; calorific value about 20MJ/kg.

MICC See *Mineral-insulated cable.*

microbore system Employs copper tubing of 6mm, 8mm and 10mm diameter, of relatively very small bore, together with suitable associated pipe fittings.
Microbore heating systems should be of the sealed type, as the small-bore pipes would otherwise easily block with scale or corrosion products.
Advantages: some cost saving, as regards the pipe system itself; reduced disturbance to decorations when installing system in an existing dwelling; no air-venting problems should be experienced with correct size pump. Microbore systems operate with high water velocities of about 1.5 m/sec. Any air bubbles in the pipes are forced through the system to the radiators, where they are vented at the air vent cock, which is usually fitted to each radiator or similar terminal unit.

microclimate The particular climatic condition of a specific (small) area, such as a building development area or site. May differ materially from (fairly) local weather station condition reports for reasons of peculiarities related to the specific area; eg mountainous terrain or surround, altitude, exposure to winds, etc. When the correct forecasting of local

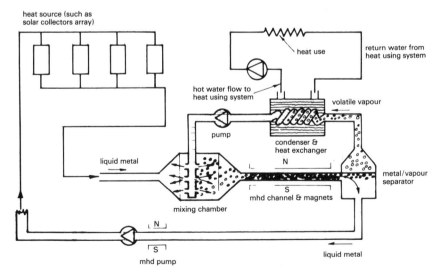

Figure 71 Exploitation of low grade heat source (solar, waste etc) using magnetohydro-dynamics principles

conditions is essential to the design of a services system (say, solar-energy collection), the microclimate of the site must be closely studied and reliance not be placed entirely on reports issued by the nearest meteorological station.

mineral fibre thermal insulation products Rock fibre slabs, mats and blankets of various densities and thicknesses bonded to produce flexible and semi-rigid materials with a low proportion of phenolic resin. Speciality faced, wired and shaped products are also available. *Application:* thermal, acoustic and fire

insulation for use in a range of temperatures up to 1,000°C (1,832°F).
Energy-saving potential: 75mm of insulation on a factory roof can bring a saving of up to 80 per cent compared with an uninsulated roof.

mineral-insulated cable (MICC) Has its conductors encased inside a metal sheath insulated with a mineral powder. An alternative wiring system to conduits, etc systems. Particularly suitable for outdoor use, in hazardous locations and in areas of heavy condensation.

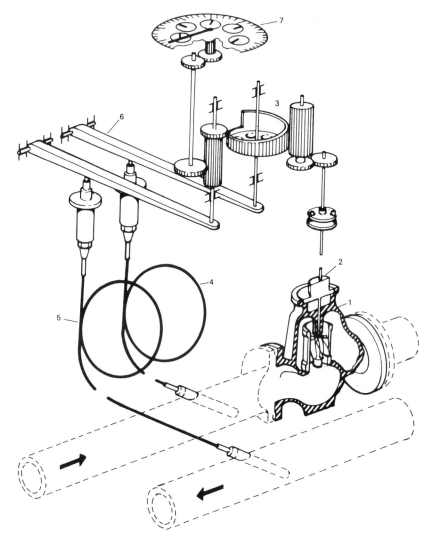

Figure 72 Aquametro heat meter

modified silica Refractory material which, by the addition of ferrous slag (plus carbon to retain the iron in the ferrous condition), lowers the temperature at which quartz inverts to tridymite and thus produces a refractory brick relatively free of unfavourable expansion characteristics. In spite of its slightly lower melting point, this modified 'black' silica brick gives better results than an ordinary silica brick.

modular boiler For the provision of central heating and hot water. Each unit is made up of a number of identical heat-exchanger modules venting into a common flue casing. Modules are typically 50kW, and units are available up to a maximum of 12 modules.
Application: for use in industrial and commercial premises, particularly where wide fluctuations in demand occur.

moisture content (wood) The water contained within a particular wood is calculated as a percentage by weight of the dry wood. The percentage does *not* relate to the total weight of wood substance and water.
Illustrative example: a piece of wood of weight 2kg at a moisture content of 100 per cent contains 1.0kg of wood substance and 1.0kg of water.

moisture content (wood) – effect of Wood waste is considered as relatively dry when the moisture content does not exceed 20 per cent (by weight); it can then be incinerated, without undue difficulty, in almost any type of furnace. Certain modern underfeed stoker type furnaces can burn wood waste with moisture contents of up to about 50 per cent, whereas wood wastes with moisture contents of over 50 per cent, 100 per cent and sometimes 150 per cent are very difficult (or impossible) to burn in such furnaces. Conventional incinerators cannot successfully combust very wet material; predrying of the waste must be achieved, unless the residence or dwell time of suspended waste matter in a furnace (such as a cyclone) allows drying before ignition within the combustion chamber. The burning of wet waste is not as efficient as the burning of dry waste because a proportion of the heat release is expended in converting the water into steam. The caloric value of wet waste/kg is considerably lower than that for dry waste.
(Figure 73 shows the variation of calorific value with moisture content for some various types of wood.) The presence of a high proportion of wet waste will always reduce the furnace output.

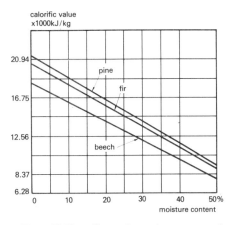

Figure 73 Chart illustrating moisture content/ calorific value variation

moisture movement Refers to the change in dimension(s) of hygroscopic materials in response to changes in moisture content. Such materials expand with an increase and contract with a decrease in moisture content in response to changes in relative humidity.

monitoring system Facility for maintaining a continuous or intermittent check or observation on the performance of an operating sevice (eg electric-power consumption, oil flow, environmental conditions, underground pipe mains).

monolithic (plastic) refractories Plastic refractories mixtures prepared under controlled conditions used for furnace repairs or complete furnace linings. Can be formulated for specific applications, such as for the refractory lining of cyclone furnaces. (See figure 74.)

motor – synchronous See *Synchronous motor.*

mud-hole Inspection opening into water side of steam or hot-water boiler provided with sealed cover.

mullite An aluminium silicate mineral – $3Al_2O_3 \cdot 2SiO_2$.
It is the only compound of alumina and silica which shows stability at high temperatures. Mullite and glass are the final products obtained by heating andalusite, cyanite and sillimanite. Mullite is one of the principal phases found in the heating of fire-bricks; it is a rare mineral, which occurs in rocks in the Isle of Mull.

multiflue chimney Has two or more separate internal compartments, usually one per connected furnace or boiler. A recommended arrangement for optimum draught management. (See figure 75 .)

multizone air-conditioning heating units
Complete heat/vent/cool packaged system capable of conditioning up to 20 zones independently. Options include heating by gas, electric, hot-water or steam coil and cooling by chilled water.
Applications: large industrial, commercial retail or institutional applications.
Energy-saving potential: current systems incor-

porate solid-state electronic control system, enthalpy control, night set back control, two-stage gas heating, staged heating, two speed compressors and multizone operation concept.

multizone system Employs an air-handling unit which has a heating battery and a cooling battery in parallel. Zone dampers are provided within the unit across the hot and cold deck at the discharge from the air handler. Separate ducts are then run from each set of dampers to each separate zone. Cold and hot air are mixed by the damper in the required proportions to satisfy the thermal demands of the particular zone served.

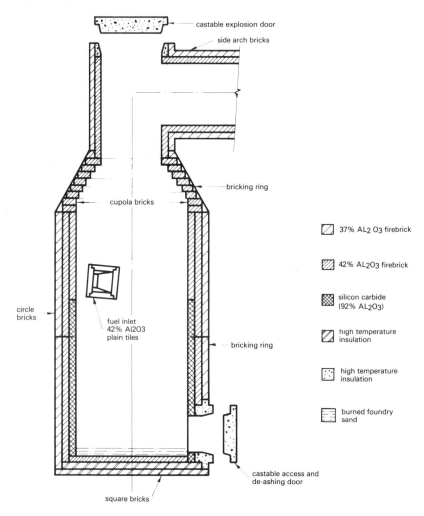

Figure 74 Cyclone refractory composition

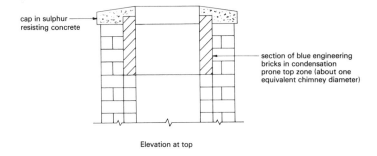

cap in sulphur resisting concrete

section of blue engineering bricks in condensation prone top zone (about one equivalent chimney diameter)

Elevation at top

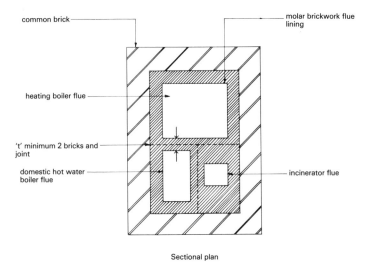

common brick

molar brickwork flue lining

heating boiler flue

't' minimum 2 bricks and joint

domestic hot water boiler flue

incinerator flue

Sectional plan

Figure 75 Multiple compartment chimney construction (above) and section through multi-compartment chimney (below)

National Coal Board Controls and operates the British state-owned (nationalised) coal-mining industry.

natural convector See *Convector – natural.*

natural draught See *Chimney draught – natural.*

natural frequency A frequency of a system or material at which it freely vibrates when a force is applied and removed (eg kicked).

natural gas Hydrocarbon gas drawn from underground sources (as opposed to manu-factured gas).
Main constituents: methane, and ethane, in conjunction with inert gases. Typical gross calorific value: $38MJ/m^3$.

NCB Abbreviation for National Coal Board.

necking and grooving Special case of corrosion in boilers which occurs mainly on smoke tubes, stay bolts (rods) and in cylindrical boilers where the tubes enter the boiler end-plates.

net long-wave radiation The net available radiation, excluding the incoming and out-going short-wave radiation.

net present value Conservation projects usually involve expenditure to achieve savings. Expenditure is regarded as negative; savings as positive (cashflow). The summation for a project of expenditure and savings over a given period is termed the net present worth. Eg, assuming an initial expenditure on a heat recovery wheel to a swimming pool complex, there is an initial expenditure followed by the accrual of energy savings (discounted to their present value). The sum of the expenditure (negative) and of the savings (positive) over a given period is the net present value.

net present value analysis (NPV) A technique for analysing the economics of systems which have a long life expectancy, involving discount-ing future economic benefits to their present value and using appropriately selected interest rates. The assumed interest rate is critical to the outcome of the analysis.

net radiation (solar collection) The difference between the total incoming and total outgoing radiation. Incoming radiation is made up of (1) short wave from sun and sky and (2) thermal, mainly from atmospheric layers close to the ground; outgoing radiation from a surface comprises (1) reflected short-wave component

depending on incident irradiance and reflec-tance of the surface, and (2) long-wave thermal radiation component depending on ground surface temperature and long-wave emittance.

neutral point Relates to pumped heating systems and denotes that point where there is neither negative nor positive pump pressure on the system.

Newton's law of cooling The quantity of heat which passes from a liquid which is cooling per unit time is proportional to the temperature difference which exists between the liquid and its surroundings.

nitrogen generator Produces in-house nitrogen and eliminates the need for purchase or hire of cylinders of the gas. Operates essentially by compressed air at minimum pressure of 7 bars passing through a carbon sieve. Separation occurs by pressure swing adsorption, in which two beds of carbon material are exposed to the compressed air in a cyclic process. Generator can be on stream at a set purity within 20 minutes of start-up. The nitrogen is usually stored in cylinders at pressure of 5 bar.

nocturnal cooling Cooling of an open surface which is exposed to the night sky. A clear sky at night can have considerable cooling effect, possibly justifying provision of surface insulation.

noise Unwanted sound.

noise criterion curves (NC) A set of curves based on the sensitivity of the human ear. They give a single figure for broad band noise. Used in the USA for indoor design criteria.

noise rating curves (NR) A set of curves based on the sensitivity of the human ear. They are used to give a single-figure rating for a broad band of frequencies. Used in Europe for interior and exterior design criteria levels. They have a greater decibel range than NC curves.

non-return valve Permits the flow of fluids in one direction only. Used to be commonly fitted around the circulating pump of a heating system to permit partial-gravity (thermo-syphon) circulation when the pump was switched off.

north light glazing insulation See *Thermal insulation – north light glazing.*

nozzle See *Air supply nozzle.*

N

off-peak electricaire heaters Industrial, multi-module off-peak electric storage, warm-air heaters. Typically multiples of 21kW on common plenum with floor-mounted fan for ducted warm air systems.
Application: off-peak heating of large industrial and commercial premises by ducted warm air (where a off-peak tariff is available).

off-peak electric heating system Operates with lower-priced off-peak electricity.

off-peak electricity Available at times of generally low demand for electricity (usually at a special reduced tariff). Generally restricted to certain periods.

oil additives See *Fuel oil additives.*

oil/ink mist filtration equipment Collects stray mist from workshop environment, allowing air to be recirculated and machine tool oil (collected in mist form) to be reused.
Application: most forms of machine tools, autolathes, screw machines, newspaper printing machinery.

Energy-saving potential: filtration of mist at source can obviate the need to exhaust air to the outside via ventilation or direct ducting and can, thereby, reduce heating costs and save oil coolant.

oil meter Permits accurate calculation of burner consumption and checking of related boiler efficiency, heat-saving methods and pipelagging improvements.

oil shale Fossil-fuel resource in the form of sedimentary rock which contains a wax-like hydrocarbon. Vast quantities of deposits found in many countries. Expensive to extract by present techniques relative to conventional oil extraction. Will become valuable source of energy in the future.

open-cycle gas turbine Takes air from the atmosphere, compresses it and heats it at constant pressure by the combustion of a fuel. The exhaust gases are discharged at very high temperatures and can be piped to produce additional thermal energy. (See figure 76.)
Application: general purposes.

O

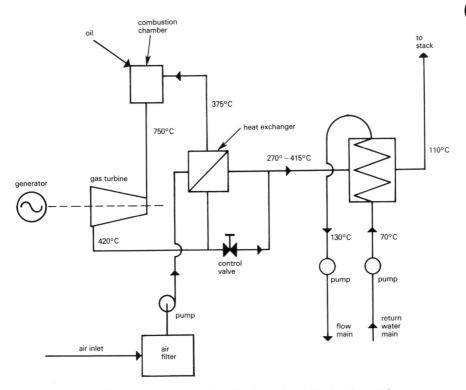

Figure 76 An open cycle gas turbine circuit used for district heating supply

125

optimiser See *Computerised optimiser.*

optimiser control Employed with space-heating systems to optimise the timing of heat input relative to external and internal environmental conditions.

optimum start control (also **optimiser**)
Operates by sensing the inside or outside temperature and controlling the space heating to ensure that the building is up to temperature by the start of the occupancy period; usually enclosed inside a plug-in module and uses electronic microcircuitry.
Applications: commercial and industrial premises requiring improved control to allow for maximum comfort at minimum cost.
Energy-saving potential: savings of 10 per cent to 35 per cent can be achieved using this type of control in commercial premises, schools, etc.

organic decay Relates to the production of biogas. The organic material from which the gas is generated is termed the feedstock. The decay process may follow one of two routes: *anaerobic* or *aerobic.* Figure 77 indicates the end products of organic decay.

organic Rankine cycle system For the generation of electricity by the evaporation and expansion of refrigerant. Whilst in principle a normal steam turbine cycle, the medium water is replaced by a refrigerant.
Applications: electricity generation from steam at atmospheric pressure, from fluids at 90°C to 150°C (194°F to 302°F) or from gases at 150°C to 400°C (302°F to 752°F). Also for power generation in conjunction with solar panels.

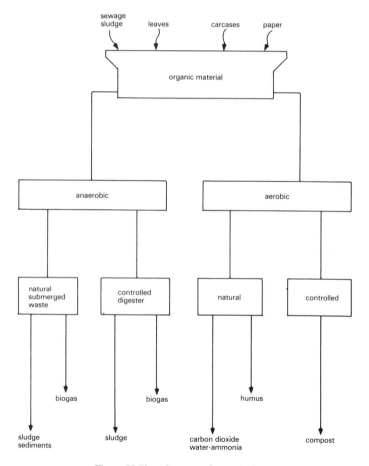

Figure 77 Flow diagram of organic decay

orientation The layout of a site, building, wall, window, etc relative to the points of the compass and to the latitude and longitude. Of special importance with respect to heat-gain and heat-loss calculations and to solar-energy collection.

Orsat apparatus Glass constructed instrument for the determination of carbon dioxide, carbon monoxide and oxygen in flue gases. Cumbersome assembly of parts; largely superceded for field use by more compact instruments. Valuable aid to teaching.

osmosis Diffusion process across a membrane placed between two different fluids which occurs because of nature's tendency to equalise concentrations. For example, when a semi-permeable membrane (permeable to water but not to salt) is placed between sea water and pure water which are both under the same pressure, diffusion of fresh water into sea water will then occur because of osmosis.

osmosis – reverse The reverse of osmosis achieved by applying pressure of about 30 bars through a semi-permeable membrane to achieve a reversal of the natural direction of the osmosis process. For example, to filter (purify) salt water, a pressure of 30 bars is applied to the sea water; it will then filter through the membrane and be converted to fresh water. (See figure 78.)
Reverse osmosis can also be applied to achieve the removal of dissolved solids in water treat-ment for district-heating and similar systems. The membrane used in reverse osmosis plants is either a hollow fibre polyamide or cellulose acetate supplied in spiral-wound sheets.
Reverse osmosis offers these advantages over conventional exchange water treatment systems: no requirement for regeneration; plants are compact and can be used on a continous basis.
Application: best used with brackish water with a high dissolved solids content. Plant is flushed out at monthly intervals with detergent or acid; the effluent (a concentration of the

solids originally present in the water) can be discharged into the public sewers.

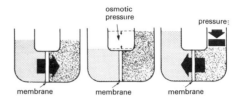

Figure 78 Principle of reverse osmosis

outside surface resistance Term used in U-value computations. Value depends on wind speed exposure.

overfire jets High velocity jets of air or steam directed over a fire bed in a furnace to promote thorough mixing of the combustion products. Steam jets now only seldom used due to inherent loss of energy.

overhead radiant heater Can be in the form of an integrated system of gas-fired overhead radiant heating with the products extracted by an induced-draught fan. Reflectors placed over the radiant pipes ensure that the heat is directed downwards to where it is most required.
Applications: for use in industrial and commercial premises where it is not practical to raise the temperature of the entire volume of air. Particularly appropriate for buildings of thin construction, poor insulation or with excessive ventilation.
Energy-saving potential: avoids necessity of long warm-up periods. Spot heating, individual adjustment as required.

oxygen sampling system Custom-built permanent or versatile trolley systems for the continuous monitoring of flue gas. The systems usually provide digital readout and are equipped with power relays to initiate an external control or alarm. Can be fitted with chart-recording facilities.
Application: for use with boilers or furnaces.

packaged boiler Preinsulated (more or less) compact matched factory-assembled combustion unit. Works tested prior to delivery to site in one 'package'. A truly packaged boiler requires merely to be placed on its foundations and to be connected to the chimney, electric supply and piped connections to become fully operational. (See figure 79.)
Such boilers are available in a wide range of duties and with different firing arrangements (oil, gas, coal, wood waste, etc).
Selection of package boilers presupposes adequate access to the boiler location. Most types of packaged boilers can be somewhat broken down into the major components to facilitate transport and access.

package deal Related to energy conservation and building services form of installation contract in which *all* the work is carried out by one contractor (as opposed to specialist contractors who will deal respectively with, say, civil engineering, pipe work, electrical installations, etc).
Advantage: single control and responsibility.
Disadvantage: the most suitable specialists may not be employed under such deal.

panel radiator See *Radiator – panel.*

panel radiator – oil-filled See *Radiator – oil-filled.*

parallel circuit or connection Electric conductors or circuits are said to be in parallel when the respective terminals are of the same polarity and are connected in such a manner that the current divides between them.

passively heated solar building A building in which solar input is stored and used internally without the aid of mechanical equipment.

payback condition (simple) Economic stipulation which requires that the achieved life of a (conservation or solar) system shall exceed its initial cost divided by the annual value of the fuel saved by installing the system, ignoring all interest charges and inflation.

payback – discounted Conservation project involving expenditure and energy savings involves a change with time of the net present value. At the outset the net present value is negative; as the savings accrue, it becomes positive. The discounted payback period is the cross-over point at which the net present value is zero.

P

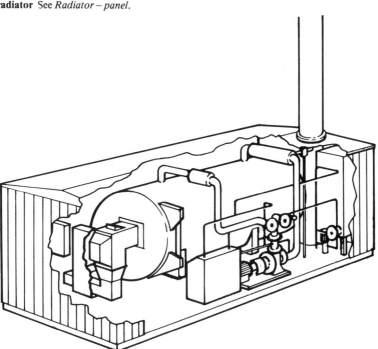

Figure 79 Diagram of a packaged heating plant

payback – optimum level Optimum level calculation determines the point at which the total projected expenditure on energy and on energy conservation is at a minimum.

payback – sensitivity analysis Directs special attention to one or more factors which are critical to the overall cost of the particular project. May involve indepth analysis of factors such as forecasting of fuel cost trends, interest rates, etc.

peat Fossil fuel formed comparatively recently. Low calorific value which mitigates against transport of fuel over long distances. Usually consumed close to source. Major users in Soviet Russia and in Ireland.

perfect gas A gas which at all conditions of temperature and pressure satisfies the relationship:

$$pv = RT$$

p denotes pressure
v denotes volume occupied by 1 mol of the gas
R denotes gas constant
T denotes absolute temperature

Oxygen, nitrogen, helium and hydrogen closely approach the requirements for a perfect gas.

perimeter heater See *Convector – continuous.*

permafrost Relates to perennially frozen earth.

permanent hardness Retained in the water after heating and boiling. It therefore does *not* cause scale deposition.

phase advancer Special machine which is connected in parallel with individual motors or other inductive apparatus, and is switched on or off with the motor. The power factor is improved in the consumers' cables and switch-gear up to the motor terminals.

phosphor bronze Alloy consisting in the main, of copper tin and phosphorus.
Application: where resistance to corrosion and wear is essential; eg fan and motor bearings, steam valves, etc.

photo-control unit Automatic switching of lighting activated by diminishing or increasing levels of natural light, independently of time-switches, etc.
Applications: control of outside lighting (factory-yards, docks, etc, public street lighting), or interior lighting (offices, shop windows, warehouses, etc).
Energy-saving potential: with an installed lighting load from 10 to 50 kW the payback period can be between 17 weeks and 20 months.

photovoltaic cell Photovoltaic cells are the key to using sunlight to generate electricity – as opposed to just heating water (such as via solar collectors). Such cells were originally pioneered by NASA for its space programme; they consist of thin wafers of silicon which are chemically treated to divide them into two layers that possess different electric potential. When the cell receives sunlight, an electric charge is generated which by the difference in potential between the two layers is separated into positive and negative. An external electric circuit then collects the electricity thus generated.
A panel of 24 cells arranged in series can produce 12 volts, sufficient to charge a battery. Fed by way of a *regulator,* it can supply direct current, or, fed through an *inverter,* alternating current. An essential feature of any system is a set of batteries to store power for use when there is no solar-energy input (eg on cloudy days).
A photovoltaic system is claimed to be roughly comparable in cost to that of diesel generators, making it suitable for supplying power to remote houses, island communities, out-of-the-way telecommunication relay stations, radio beacons and so on – all cases in which it would usually be far more expensive to install an electric mains supply.
A photovoltaic system has the major advantage over diesel generators that it requires very little routine maintenance – no more than periodically dusting the cells – especially beneficial in developing countries, where maintenance facilities are usually poor and where many an expensive telecommunications system lies idle for want of a spare part or a gallon of diesel oil at a remote hilltop relay station.
At present, photovoltaic systems are a very expensive way of producing electricity. At peak output, it currently costs $10 to produce 1 watt. Electricity produced by either nuclear or conventional thermal power stations costs in the region of $4 per watt. Though the cost of photovoltaics has already fallen sharply from $50 per watt in 1975, the current main effort is aimed at reducing the cost to about half its present level.
The most costly single item in a photovoltaic generator is the polycrystal silicon, accounting for about 45 per cent of its cost. Photovoltaic companies are working on the development of a less pure and cheaper, but equally effective, type of silicon – known as amorphous silicon – which could substantially reduce the cost.

(See figure 80.)

Applications: the world's largest power pilot plant – located in the Mojave Desert of California, USA – (designed to generate 10,000 kW) is now nearing completion. It consists of a 91.4m diameter central receiver solar thermal power system (the power tower), which is surrounded by some 1,800 heliostats; these are giant sized movable mirrors which reflect the incident solar energy onto the power tower to generate steam for supply to the turbine electricity generators.

A large operational photovoltaic power station is located 50km north-west of Riyadh, Saudi Arabia, and provides each household in two villages with a public electricity supply. The station and its solar-energy receiving field of 160 photovoltaic array panels, housing 40,960 solar cells, converts sunlight into electricity with an output of 350 kW of power. The panels (which rotate horizontally and vertically to track the movement of the sun) cover the enormous area of about 293,000m². The solar power feeds electricity into the grid system and also to a 1100 kW battery system, from which electricity is drawn during the hours of darkness. Stand-by power is available from a 1000 kW diesel generator. The cost of this project is given at $25 million.

pH test kit Used for periodic monitoring of the acidity or alkalinity of water. Usually comprises a small comparator: a few drops of the testing solution are introduced into the water sample, and the ensuing colour is compared with standard slides mounted on a disc. Most liquids used for pH measurement are effective over only a narrow range, so that the suitability of the test liquid must be established for each particular application; once the required range is known, one or two test liquids are likely to cover the range of operations.

pH value A measure of the degree of acidity or alkalinity of water related to the concentration of hydrogen ions. In water of absolute purity at 21°C (70°F) the concentrations of hydrogen and hydroxyl ions are equal, and each may be expressed as 10^7 ions per litre. The pH scale uses the logarithm to base 10 of this value and changes the sign; thus the hydrogen ion content of pure water at 21°C (70°F) is stated as pH 7, the neutral condition. Water having a deficit of hydrogen ions (pH 7) is alkaline, and water having an excess of hydrogen ions (pH 7) is acidic.

pigging Technique commonly employed for separating products in, or cleaning, energy pipe lines.

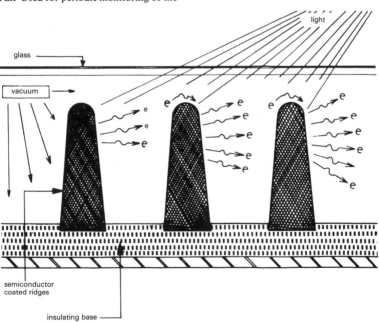

Figure 80 Principle of micro-corduroy (thin film) photovoltaic cell for electricity generation

pig trap Access arrangement to transmission system for introduction or retrieval of piped pig travelling device.

pilot burner Gas (or other fuel) separate burner section within a firing system which maintains a constant pilot flame from which the main burners (jets) are lit in response to heat demand. Pilot burner pipe connection is usually fitted with a separate isolating valve or cock.

Since this burner remains in use continuously, it must be carefully adjusted (with instruments) to maintain a flame of minimum output related only to the requirement of the flame remaining stable and being adequate for lighting the main jets.

The most modern boilers tend to dispense with pilot burners and employ electric or electronic main jets ignition methods.

See also *Pilot flame.*

pilot flame Relates mainly to gas-fired boilers, heaters and process equipment. The flame is maintained in continuous operation, and the main gas jets are lit from this flame on heat demand.

Caution: flame must be stable to withstand the disruptive effect of draughts.

pilot light Relates to gas burners and gas-fired furnaces. A small jet which is permanently alight (whether heat input is required or not) and has the purpose of smoothly lighting the main gas jet(s) when the controls call for heat input. Correct adjustment of the pilot flame is essential to minimise gas consumption.

Modern appliances tend to incorporate electric ignition means to obviate the pilot light gas wastage.

pipe anchor Fixed point (structure or special steel or concrete post) from which a compensated pipeline will expand or contract, there being no movement at the anchor point.

piped pig – electric transmission system Technique of pigging adapted by the CEGB to minimise the risk of underground electricity cables overheating under load. An interesting application of such method is employed as part of a scheme of transferring electric power from generating stations on the Thames and Medway estuaries into central London; two three-phase 400kV cable circuits are being installed from an overhead line at West Ham to a sub-station at St John's Wood. For most of the 16km route, the cables are laid in concrete troughs constructed in towpaths of the River Lea and the Regents Canal.

To keep the cables cool, four 75mm nominal-bore polythene water pipes are laid with the cables in the trough, which is then backfilled with cement-bound sand. Water is taken from the canal at several points, pumped through the water pipes and discharged into the canal about 2.5km away.

To prevent build-up of microbilogical growth and debris, which may impede the water flow, and also to remove trapped air, pigs will be passed through the pipes at regular intervals; these will be 125mm diameter polyurethane foam units, which will pass through the 75mm pipes when compressed.

They will be launched into and retrieved from the cooling-water network by a system of purpose designed pig traps. Forty-eight launching traps and eight receiving traps will be integrated into the network.

piped pig – gas transmission system A travelling device, termed a 'pig', which has been developed by British Gas and is widely used for inspection and remedial work related to gas transmission pipes.

pipe fitting – capilliary See *Capillary pipe fitting.*

pipe fitting – compression See *Compression pipe fitting.*

pipe freezing Process of pipe isolation by applying freezing jackets to each side of a section to be frozen with the object of producing an ice plug which will isolate the section (like a closed valve). Freezing jackets operate with injection of carbon dioxide. Freezing operation avoids need to drain down a complete water-carrying system when work is required on only a section of it.

pipe header See *Header.*

pipe-in-pipe Term generally applied only to an underground pipe assembly which consists of an insulated (thermally) service pipe or pipes encased in a pressure-tight casing of a suitable material – may incorporate an air gap between the thermal insulation and the outer protective casing.

pipeline identification Applied to multiservice pipe systems to facilitate the speedy identific-ation of the service provided by a particular pipeline. Colours are specified by B.S. 4800 as under:

Water services basic pipeline (insulation): green (12D 45)

drinking water: auxiliary blue

132

cooling (primary) water: white
boiler-feed water: white between crimson
condensate: green between crimson
chilled water: green between white
low-temperature central heating: crimson
  between auxiliary blue
high-temperature central heating: auxiliary
  blue between crimson
cold down-service: auxiliary blue between
  white
hot-water supply: crimson between white
hydraulic power: salmon pink
sea or river water (untreated): green
fire-fighting service: red
Gas basic colour: yellow ochre (08 C 35)
  manufactured (town) gas: green
  natural gas: yellow
Oils basic colour: brown (06 C 39)
  diesel fuel: white
  boiler and furnace fuel: brown
  lubricating oil: green
  hydraulic power: salmon pink
  transformer oil: crimson
Sundry services
  compressed air: light blue (20 E 51)
  vacuum: white on light blue
  steam: silver grey (10 A 03)
  drainage: black (black)
  electrical conduits and ducts: orange
    (06 E 51)
  acids and alkalis: violet (22 C 37)
Note: B.S. 4800 provides a numbered code for
each of the above *basic* colours.
The identification may be applied by painting
coloured strips on to the surface of the service
pipes or, more commonly, by the application
of adhesive colour strips to the surface of the
pipes.

pipelines Pipes joining service points some
considerable distance apart. Smaller branch
pipes may be connected to the pipe lines.

pipe sleeve See *Sleeve.*

pipe support – glide type See *Glide pipe
support.*

pipe support – spring type See *Spring pipe
support.*

piping complex Multiple assembly or network
of pipes associated with major engineering
installations, such as in factories, district
heating, refineries.

piping system May consist of pipelines and/or
pipe complexes.

pitot-tube Portable apparatus commonly used

for field measurement of air velocities and
pressures. Measures the velocity head from
which the velocity in the immediate vicinity of
the instrument can be calculated. In use, the
tube is inserted into the duct or pipe facing the
direction of flow.

pitting Related to steam boilers; a major form
of corrosion encountered on the water side of
boilers. Most common where feed water has a
pH value of between 6 and 9 and is associated
with the diffusion of oxygen through semi-
stable protective films. Leads to 'air bubble',
'scab' and/or 'soft-scab' pitting.

plate heat exchanger Typically comprises a
pack of metal plates in corrosion-resistant
material enclosed in a frame. Liquids flow in
thin streams between the plates, while periph-
eral gaskets control the flow. Troughs pressed
in the plates can give extreme turbulence;
together with a large surface area produce high
heat transfer.
Applications: heat recovery; heating, cooling,
pasteurising, sterilising.

plate heat recovery unit In one typical such
unit, streams of warm exhaust air and cool
fresh air are finely spread and caused to flow
past each other by separating aluminium
plates. The streams do not come into contact
with each other. The aluminium plates between
the air streams extract heat from the exhaust
air and transfer it to the incoming fresh air.
Due to the separation, impurities, odours,
dampness, etc, do not communicate with the
incoming air.
Application: heat recovery from exhaust air
and gases. Typical units can handle air volumes
from 500m³/hr to 60,000m³/hr at temperatures
up to 82°C (180°F).
Energy-saving potential: formal tests on one
type of such equipment have confirmed heat
recovery efficiencies of between 60 per cent
(commercial buildings) to 85 per cent
(swimming pool halls).

plenum chamber Space *into* which input air is
fed and from where the air is distributed in a
controlled manner into the ventilated or air-
conditioned environment. May be formed as
space between suspended ceiling and building
soffit or be a purpose-made box.

pneumatic control Operates by the application
of compressed air. The controlled equipment
incorporates spring-loaded diaphragms to
which varying air pressure is applied to open,
close or provide intermediate setting of equipment such as valves and dampers. Actuating

devices can be thermostats, humidistats, pressure stats, etc. (See figure 81.)

Pneumatic controls are in competition with electric and electronic controls; whilst these are connected by electric wiring, the pneumatic control system is joined with small bore tubing (copper or plastic). A choice of systems is usually based on an assessment of relative cost. Pneumatic systems tend to be cheaper for the very large control applications.

Technical advantages are inherent in the pneumatic method for certain processes; the performance of the controlled equipment can be changed by substitution of different springs. Maintenance of a pneumatic system is considered simpler than that of a complicated electrical/electronic installation due to the basic simplicity of the controls. However, the cost of maintenance can be considerable, as the springs tend to have a limited life.

Good maintenance access must be provided to all system components.

The installation must incorporate suitable means for draining water from the system, and such drainage operation must be a routine maintenance function.

pneumatic controls Operate with compressed air.

pneumatic controls – coal delivery Discharges coal from the supply vehicle to the store or bunker in a pipe under pressure. In NCB area, pipe is of 127mm diameter and terminates with a coupling close to the delivery vehicle standing area. The delivery pipe should be as straight as possible, with only a few (necessary) bends. The final exit velocity of the coal into the store must be reduced by enlarging the delivery pipe (in NCB area from 127mm to 178mm) in its final leg at a point about 2.2m from the bunker entry. When discharging into a bunker, the displaced air must be safely vented and filtered. Pneumatic delivery is convenient and attracts, therefore, a delivery surcharge.

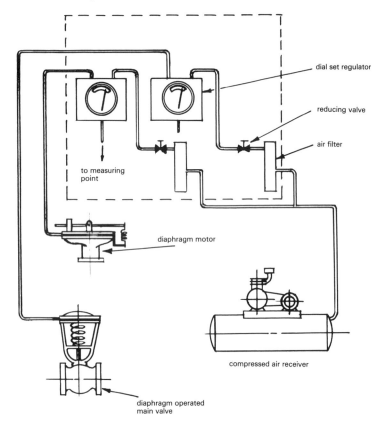

dial set regulator

reducing valve

air filter

to measuring point

diaphragm motor

compressed air receiver

diaphragm operated main valve

Figure 81 Pneumatic control system – diagrammatic arrangement

polluted air controller Monitors cigar and cigarette smoke in the atmosphere and switches on an extractor fan when a predetermined contamination level is reached, switching it off when the room is clear. (The controller can be switched to an alarm condition to monitor carbon monoxide, methane and hydrogen.)

polypropylene ball blanket For covering tanks, baths, etc to minimise fumes, retain heat etc.
Applications: boiler-feed condensate return tanks, heated processing tanks, heated liquid-holding tanks, plating baths.
Energy-saving potential: use of the spheres offers reduction in fumes and odour, reduced maintenance and evaporation losses combined with complete and immediate access to processing liquids. Energy savings up to 69.9 per cent (liquids maintained at 98°C (208°F)) have been demonstrated.

polyurethane foam Rigid polyurethane foam sprayed *in situ* internally or externally on roofs or walls.
Applications: agricultural buildings and factories; flat roofs and corrugated metal walls for purpose of thermal insulation.

porosity The amount of pore space in a material is a measure of its porosity. The volume of pore space per unit volume of material usually indicates the percentage porosity. The true porosity is a measure of the total pores, both open and sealed; a measure of open pores is only the apparent porosity.

pot burner See *Vapo(u)rising Burner*.

potential energy See *Energy – potential*.

potentiometer Instrument which compares the electromotive forces and potential differences in an electric circuit by the null method (ie does not depend on measuring deflection). Incorporated in detecting and control equipment.

power factor Relates to the non-inductive electrical loading of an electric network. The greater the inductive load (eg induction motors), the lower the power factor.
The majority of industrial loads have a power factor of less than unity; in most cases where induction motors predominate, the power factor will be in the order of 0.7 lagging.
A low power factor is undesirable, as it reduces the effective current carrying capacity of a distribution network. Most supply companies penalise the consumer with a low power factor under the terms of the supply tariff agreement.

power factor correction equipment Function to convert a lagging power factor to, say, an overall factor of 0.95. The necessary *leading* current may be supplied by employing phase advancers, synchronous motors (running over-excited) or capacitors.
Application: electrical network which operate at low power factor with the object of achieving better current loading of the electrical distribution equipment and a consequent reduction in electrical charges.
Energy-saving potential: can be very considerable and depends on the proportion of the inductive loads served off the network.

pre-filter air filter See *Air filter – pre-filter*.

pre-insulated pipe system Consists of metal (steel copper, stainless steel, etc) pipe which has been provided with comprehensive thermal insulation under secure factory conditions and inspection prior to delivery to the site of the works in specified (or standard) lengths. Insulation usually consists of polyurethane foamed material within which the pipe may move freely or to which the pipe may be firmly bonded. (See figures 82, 83 and 84.)
Such systems incorporate heat-shrunk waterproof joints, building entry arrangements, expansion and anchoring provision, valve pits, drain pits and alarm/monitoring cables.
Wide application in modern district-heating networks. Offer the major advantages of reduced site labour costs and of high standards of factory production and quality control.
Costs of pre-insulated district-heating networks are considered to be about two-thirds those of duct-located equivalent installations and offer less risk of failure due to water penetration to the pipes and insulation. Most systems are suitable for a maximum operating temperature of 130°C (266°F).

pre-insulated pipe system – bonded The thermal insulation is firmly bonded to the pipes, so that the pipes and insulation move together with expansion and contraction. This imposes greater stress on the pipeline system and may lead to failure if the yield stress of the material is exceeded. Limits the length of pipeline between anchor and expansion facility.

pre-insulated pipe system – expansion bellows Incorporated within the insulation system within the outside casing of the pipe. They do not require location within a special expansion chamber and thus eliminate a serious potential failure risk.

pre-insulated pipe system – valvepit

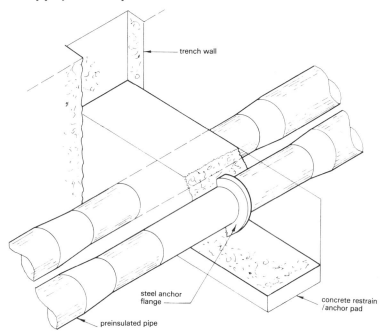

trench wall

steel anchor
flange

concrete restrain
/anchor pad

preinsulated pipe

Figure 82 Pre-insulated pipe system anchor point

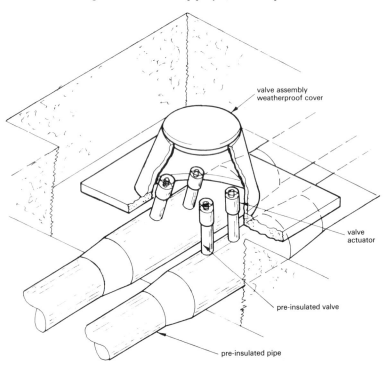

valve assembly
weatherproof cover

valve
actuator

pre-insulated valve

pre-insulated pipe

Figure 83 Pre-insulated pipe system – valve pit

136

pre-insulated pipe system – free-moving Incorporates a coat of bitumen between the steel pipe and the insulation which permits the pipe to move freely *within* the insulated section. Such networks operate with less stress than bonded ones. In the event of exposure of the pipe due to damage, the bitumen coating acts as a corrosion inhibitor.

present worth A sum X, expressed in monetary terms, to be paid in the future will be worth less than its present monetary value. Invested now, it would earn interest. To establish the true value of X in today's terms, one calculates it on the basis of present worth plus interest using the compound interest formula in reverse.

pressure differentials measuring equipment for commissioning of heating and chilled-water systems Portable test set. Comprises single-column type manometer connected to measure the pressure difference across an orifice balancing valve, complete with by-pass, isolating valves and lengths of connecting plastic tubing which terminate with probe units.

Most heating and air-conditioning applications operate with velocities which result in a low

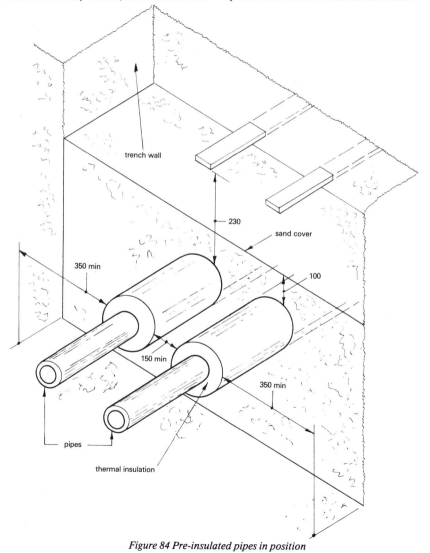

Figure 84 Pre-insulated pipes in position

energy loss, so that the pressure loss across the orifice balancing valves will require differential measurements of up to 450mm water head loss. Higher pipeline velocities require measuring devices reading up to 3,000mm water head loss and use mercury-filled test sets. It is easiest to read head loss signals up to 400mm water on a fluorescent filled manometer (specific gravity 1.88), rather than on a mercury-filled set. Test set must be used in the upright position.

pressure filter Contained in a steel pressure vessel and can operate under pressure if hydraulic conditions in the system require it to do so. (See figure 85.)

pressure jet burner Oil or gas burner which incorporates a fan and supplies the combustion air under pressure. Does not rely on the chimney draught for the induction of combustion air, but requires a suitable chimney or exhaust arrangement for evacuation of the combustion products.

pressure-reducing valve Enables point-of-use

steam (lower pressure) to be provided as variously required from a single boiler operating at its optimum generating pressure. Under dead-load conditions the valve should give an absolute tight shut-off.
Application: main application is pressure reduction of steam in, for example, heating systems, steam injection and humidifier process control.

pressure regulators – compressed air Reduce the input compressed air pressure to the required operating level for efficient operation of connected equipment.
Applications: compressed air systems; control of pneumatic tools and devices.
Energy-saving potential: a saving on total operating cost is offered through the operation of pneumatic equipment at its optimum operating pressure by reducing wear caused by excessive pressure (without significant increase in output) and by avoiding the waste of compressed air.

pressure set Offers the storage of a specified

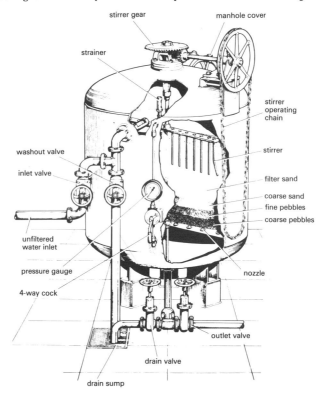

Figure 85 Cut-away view of industrial sand filter

volume of water under a cushion of air to enable a variable demand for water to be met economically.

Typically comprises two or more centrifugal pumps and a pressure tank suitably pipe interconnected, pressure switches, non-return and sluice valves, all mounted on to a common base plate, together with a control panel.
Application: boosting cold-water supplies to domestic, industrial, municipal, agricultural and marine installations.

pressure stat Switching device which is actuated by change in pressure.
Application: steam, compressed air, etc systems.

pressure switch Senses pressure differentials; offers advantage over flow switch; is without moving parts. Activates controlled equipment when specified differential has been achieved, lost or exceeded.
Applications: process control, fan-diluted gas boiler systems, etc.

pressurisation unit Purpose-made to pressurise (usually by nitrogen) water systems (hot or cold) to provide an artificial head for purpose of preventing boiling in a high-temperature (high-pressure) hot-water system or to supplement the static pressure in a cold-water distribution.

primary air As applied to a bed-type furnace, is introduced *below* the bed of solid fuel to provide the main or primary air for combustion. In liquid or gas fuel burning, the primary air is introduced into the combustion chamber with the fuel, often in a premixed fashion. In a cyclone or vortex furnace, the primary air is often the only source of combustion air and is also used to pneumatically convey the solid fuel material into the combustion chamber.

primary air supply See *Chimney – secondary and primary air supply.*

primary circuit Directly connected to the main energy source, eg boiler, chiller, etc.
Application: flow and return circulation of an indirect hot-water cylinder connected to boiler tappings or circuit.

priming Term related to steam boiler plant. Discharge of steam containing excessive quantities of water in suspension because of violent ebullition in boiler. Induced by surging and aided by high water level. Vigorous movement of steam bubbles to the surface and their explosion causes surging and artificially raises the water level. The ensuing spray tends to be carried over with the steam flow from the boiler. Priming can be avoided by limiting the concentration of dissolved solids to the recommended levels related to the particular boiler type and steam discharge pressure.

programmer – heating and hot water A more or less sophisticated time clock, arranged to cater for a variety of time and heating/hot-water functions; eg 'hot water only'; 'heating only'; 'heating and hot water', with several switching operations in a 24-hour period. Activates electric switches or valves to achieve the desired programme.

propellor fan Simplest form of fan. Commonly comprises a sheet-metal impeller with relatively large clearance inside an orifice. The fan blades – usually four or six – are bolted to a cast-iron or aluminium impeller hub. Performance depends on shape of blades: broad-bladed fan will handle more air more quietly than narrow-bladed fan of same diameter running at same speed. Air enters the fan from all directions and is discharged mainly axially. When the air meets resistance, it tends to flow backwards through the impeller.
Application: where large air volumes have to be moved against negligible resistance. Fans may be ring or diaphragm mounted. Suitable for operation in either direction. Fan total efficiency low – 60 per cent to 75 per cent for the larger sizes.

proportional band The range of values of the controlled variable which corresponds to the full operating range of the final control element in proportional-position control. Usually expressed as percentage of the full-scale range of the controller; the adjusting dial is calibrated as a percentage.

proportional control The final control element assumes a definite position for each value of the controlled variable.

psychrometric chart The graphic presentation of hygrometric data on the properties of air; it is universally used by air-conditioning specialists, as it presents much essential information very simply. Each chart is only strictly correct for the particular atmospheric pressure for which it has been prepared.
Figures 86, 87 and 88 show applications of the use of a psychrometric chart.
The chart is usually plotted with the co-ordinates of specific humidity and dry-bulb temperature (as indicated in Figure 86). The

psychrometric table

construction of such a chart is based on the following:
  lines of constant dry-bulb temperature;
  lines of constant humidity ratio;
  lines of constant relative humidity;
  lines of constant specific volume;
  lines of constant wet-bulb temperature;
  lines of constant enthalpy;
  lines of constant dew point temperature.
Any one air condition is represented by a particular point on the psychrometric chart, and this condition can be ascertained once *two* independent properties of the air (eg dry-bulb temperature and relative humidity) are known. As each property is represented by a line on the chart, the intersection of the two relevant lines establishes *that* point which indicates the condition of the air. It is stressed that the two reference conditions of the air must be truly *independent* ones as represented by the lines listed above. (See also figures 87 and 88.)

psychrometric table See *Hygrometric table.*

pump – belt-driven See *Belt-driven centrifugal circulating pump.*

pump – canned-rotor See *Canned rotor circulating pump.*

pump – centrifugal See *Centrifugal pump.*

pump – direct-coupled See *Direct-coupled centrifugal circulating pump.*

pumped storage Energy conservation method which uses low-cost off-peak energy to pump (elevate) water to a high level and allows it to fall back to its original low level to generate electricity during peak time. Applicable particularly to hydro-electric and tidal electricity generation schemes.

pump duty Specifies the quantity (1/sec or gal/min) of water (or other liquid) handled by a particular pump and the resistance (kN/m^2 or ft wg) against which it will deliver this output.

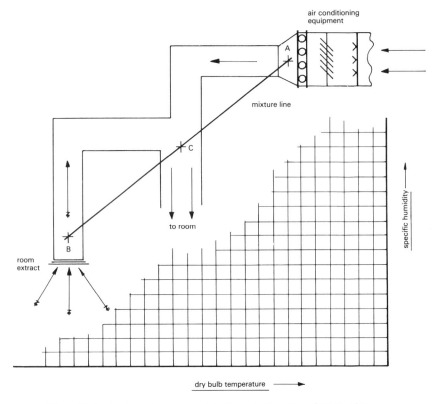

Figure 86 Application of psychrometric chart construction of mixture line

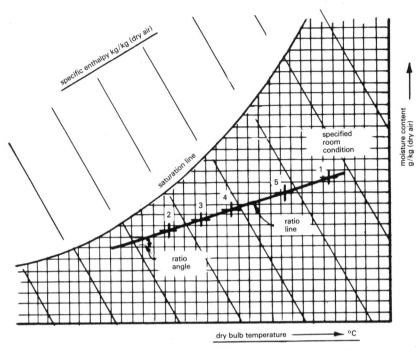

Figure 87 Application of psychrometric chart – ratio line construction

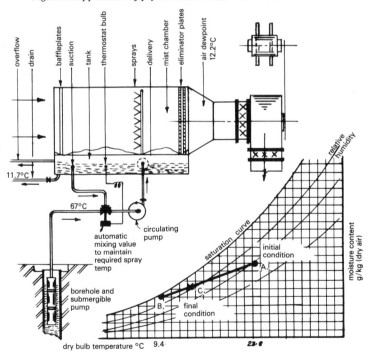

Figure 88 Application of psychrometric chart to air conditioning

pump head The total resistance external to the pump against which the specified quantity of liquid is moved (kN/m^2 or ft head wg).

pump impeller Paddle-like element of centrifugal pump which rotates inside pump casing and generates the pumping action. In the case of certain types of pumps, the performance can be varied by fitting a different impeller (within the pump characteristic), though this may also require a change of electric motor.

pump performance characteristic Usually presented in the form of curves which relate, for a particular pump, the quantity of water (or other liquid) handled against the corresponding pump head. Commonly, the X-axis indicates the flow (1/sec or gal/min) and the Y-axis the pump head (kN/m^2 or ft wg) – some graphs indicate both the Imperial and the S.I. units on opposite axes. The curves denote the *range* of duty within which the pump operates at its design condition when the pump shaft rotates at a specified speed.

pump – reciprocating See *Reciprocating pump.*

pump speed A direct-coupled pump runs at the synchronous speed of the (a.c.) electric motor to which it is shaft-connected. Belt-driven pump is provided with a pulley drive; the speed then depends on the ratio of the pulley diameters and the speed of the driving motor.

pump – split-casing Arranged to permit access to the internal parts of the pump without need to disturb the pipe connections. The procedure would be:
isolate electric supply to pump motor;
close suction and discharge isolating valves;
disconnect the wires (cables) from the motor terminal block;
disconnect electric circuit from motor terminal block;
remove the hexagon set screws which secure the motor base frame to the subframe (commonly four screws);
remove the casing nuts and, by tightening the two casing removal set screws, slide the motor and the pump internal parts out of the casing, thereby exposing the pump impeller and the mechanical seal for inspection and/or repair.

pump – submersible Constructed specifically for safe and efficient operation whilst submerged in water. May be a portable type as used for drainage during building construction or be for permanent use inside a borehole or sump.

Usually such pumps have a casing made of special cast iron, an impeller of hardened and highly abrasion-resisting casting, a double mechanical shaft seal between pump and motor, oil-filled motor, motor casing of aluminium or cast iron, bearings adapted to long-life lubrication. Pump control can be by means of float or electrode arrangement as best suited to the circumstances.
Lifting facilities must be available at site to permit the removal for inspection and maintenance of the larger and heavier submersible pumps.
The pumps operate virtually without a suction head and can be designed for pumping against considerable resistance and high lift.

pump – sump See *Sump pump.*

pump – twin set Arrangement of two pumps in parallel allied to a factory-assembled piping which terminates with one set of suction and delivery flanges for connection to the external pipework.
Advantage: neat and compact duplicate pump set made to predetermined dimensions and factory quality control.
One pump acts as 'duty', and the other as 'standby' unit.

PVC strip doors and windows Reduce heat loss through doorways whilst still allowing free access and visibility.
Applications: industrial building doorways with pedestrian and vehicular movement; division of factory areas to reduce heat loss.
Energy-saving potential: PVC strip doors have halved heating bills in some installations, giving payback within one heating season.

PWM (pulse-width-modulated invertor) Controls both voltage and frequency of a motor in the output stage, thus requiring a simple bridge rectifier in the input stage. This type of drive is usually the most expensive system, but offers the following two major advantages, particularly in respect of the variable speed control of fan and pump drives:
1. given an already existing drive, the inverter can be inserted directly between the isolator and the motor, permitting the original motor to be used (particularly if this was somewhat oversized for the application), so that only the inverter need to be purchased;
2. should the inverter fail in service, the motor can be switched directly on to the mains with the equipment through-put controlled mechanically (by valve or damper).
Typical payback available from fitting an inverter instead of using damper or valve

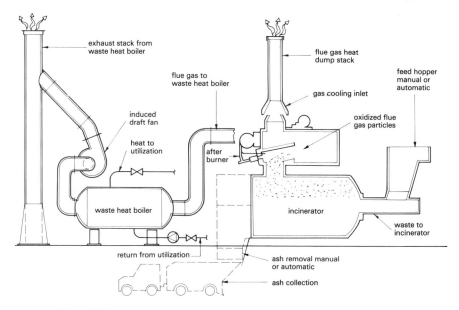

Figure 89 Diagrammatic arrangement of consumat pyrolisis unit – single module

control is in the order of two years, if the original drive can be retained.

pyranometer Instrument used for measuring the solar irradiance including both direct and diffuse components. If mounted horizontally, it will measure global irradiance. The instrument must be firmly secured in position at a standard distance from the object or surface being monitored.

pyrolisis Process which relies on the physical and chemical decomposition of organic matter under the influence of heat in atmosphere deficient in oxygen. Achieved by heating the material in a retort at high temperature in the absence of adequate oxygen for complete combustion. Material can be converted into liquid or gaseous fuels; this may be stored for use when required or be fired in a waste-heat boiler associated with the pyrolisis retort.

Storage possibility offers a major advantage over incineration. Pyrolisis operates with minimum atmospheric emissions. (See figure 89.)
Application: destruction of difficult wastes, such as tyres and plastics; operated on a large scale, can provide suitable fuel for an associated gas turbine.

pyrometer For temperature measurement above 800°C (1,472°F) for industrial or laboratory use.
Application: measurement of refractory and ferrous metal temperatures in inaccessible places.
Energy-saving potential: typically allows detection of hot spots before plant shutdown stage has been reached. It can eliminate guess-work in measuring temperatures of moving targets or targets (eg porcelain) which might be damaged by contact.

quantometer Gas meter of the turbine-operated type provided with removable, replaceable measurement cartridge. Typically has range of between 0.5 to 4,500 litre/sec. Readout can be metric or Imperial. Installation can be in vertical or horizontal pipe with or without flanges.
Application: industrial gas systems for use as secondary or check meters.

quarl Refractory throat through which a burner fires into a combustion chamber. Serves to direct the flow of hot gases and to protect the boiler at the entry point.

quartzite A refractory material which sometimes occurs as a laminated rock that can easily be split and made into blocks having much the same chemical properties as silica. It shows considerable expansion when partially converted to cristobalite and tridymite, and for this reason the blocks are usually laid up with thick raw fire-clay joints, which, because of clay shrinkage, permit free expansion of the blocks. These refractory blocks are sometimes used for lining cupolas, etc.

Q

radiant panels See *Ceiling heating – radiant panels.*

radiant strips See *Ceiling heating – radiant strips.*

radiation Transfer of heat from the heat source to other bodies remote from it without raising the temperature of the intervening space.

radiation – diffuse Solar radiation which is scattered by particles of (foreign) matter in the atmosphere and appears to come from the entire sky (as would be experienced on a hazy day or with an overcast sky).

radiation – direct The radiation from the sun which falls on a plane of stated orientation over a specified period, received from a narrow solid-angle centered on the sun's direction.

radiation – incident The combined diffuse and direct radiation components calculated proportionately to the fraction of sky hemisphere to which the plane is exposed and also calculated vectorially.

radiation – solar collection The radiant energy falling on a plane per unit area integrated over a stated period (day, month, year, etc). Normally stated in megajoules/m^2 (mj/m^2) over the stated period.

radiator Term for space-heating heat emitter. Does not describe the equipment truthfully, as heat emission from radiators is by radiation *and* convection; the proportion depends on the particular radiator design.

radiator – Column Heat emitter is assembled from a number of column-like steel or cast-iron units to provide the specified heat output. Provide a greater heat output in a given space than panel radiators.

radiator – convective Usually comprises a flat front (for enhanced appearance and radiation) and a fluted convecting section attached to the back of the flat front.

radiator – hospital Special pattern of cast-iron column radiator, designed to offer smooth surfaces and wider gaps between the columns to facilitate cleaning.

radiator – oil-filled Electric heat emitter in the form of steel or cast-iron panel or columns which is filled with oil and heated by an electric immersion heater fitted in the bottom horizontal waterways.

radiator – panel Flat heat emitter; usually manufactured of steel in single surface or multipanel configuration (one panel behind the other) to a maximum of four panels. Also available in cast-iron sections and in copper; the former are found in older installations and embodied with certain models of towel rails; copper radiators are suitable for use with direct hot water.
The design and construction of panel radiators (with the above exception) permits their use only with indirect heating systems.

radiator reflective foil Placed on walls behind radiators to reduce heat loss.
Application: hot-water heating systems utilising radiators.

radiography X-ray method of inspecting plant construction to check soundness of joints, particularly of welded joints. Widely used in the construction and insurance testing of new underground mains systems.

rapid gravity filter Removes the finer particles carried over from the settling basins. It works at high rates and can be washed easily and quickly. It is now widely adopted where filtration is necessary.

rapid recovery calorifier Incorporates (for its size and capacity) much more heat-transfer surface than a conventional hot-water calorifier (indirect cylinder); hence can provide hot water more rapidly when served off appropriately sized boiler.

rapid steam generator Typically capable of producing saturated steam at pressures up to 10 atmos in about 3 min after switch on. Available in stationary or mobile form, usually fired by 35 sec gas oil or gas. Standard fitments include steam safety valve, lack-of-water unit, excess steam pressure switch, steam temperature cut out.

raw water Water in its untreated state as it enters the treatment plant.

RDF Refuse-derived fuel, particularly as supplied by waste-recycling (sorting) plants. (See figure 90.)

reactance The *virtual* resistance of an electric circuit to an alternating current, related to the inductive and capacitive components circuits,.

R

147

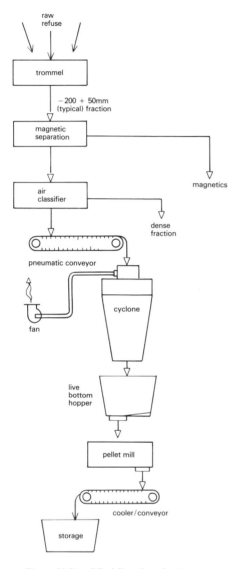

Figure 90 Simplified flowsheet for the production of refuse-derived fuel pellets

reactive attenuator An attenuator in which the noise reduction is brought about typically by changes in cross section, chambers and baffle sections; eg car exhaust silencer.

reciprocating pump Piston operated; may be driven by electric motor or by steam. Can pump against high resistance and is used widely in the feed connection of steam boilers. Steam-driven pumps must be most carefully maintained and serviced to avoid major steam wastage.

recording calorimeter Records the calorific value of burnable gases and can be used as calorific value controllers or for telemetered applications.
Application: specific and general industrial applications to ensure the effective use of bio-gases and waste-gas utilisation.

recording wattmeter system Self-contained chart recorder and transducer which provides graphic readout of power consumption in kilo-watts.
Application: monitoring power consumption in connection with maximum demand tariff, etc.

rectifier Converts alternating current into a unidirectional current; has no moving parts.

recuperative burners Recuperative tubes placed in the exhaust of intermittent kilns to preheat combustion air to a maximum of 450°C (742°F). The burner system is designed to maintain control of the air-gas ratio despite the expansion of heated air.
Application: gas-fired intermittent kilns for firing ceramics.
Energy-saving potential: fuel savings of 25 per cent have been achieved compared to similar kilns using cold-air burners.

recuperator – Escher See *Escher recuperator.*

recycling plant – mechanical sorting Operates to mechanically sort reusable components of refuse and divert same to recycling. Such plants in the USA, Italy and the UK operate on a relatively large scale, recycling glass, plastics, paper, metals from refuse collected by the municipalities.
Another by-product can be waste- or refuse-derived fuel (RDF).
Some plants use the final residue for the preparation of compost and in the production of heat or steam via incinerators.
One major advantage of recycling plants from the environmental point of view is that they *need not* incinerate at site and can export the RDF to power stations or large commercial users of boilers. This avoids the pollution associated with incineration at the recycling plant.
Figure 91 shows the general arrangement of the major recycling complex at St Louis, Missouri (USA). Figure 92 shows the plant at the Monroe County Facility and Figure 93 that at the Rome Recycling complex.

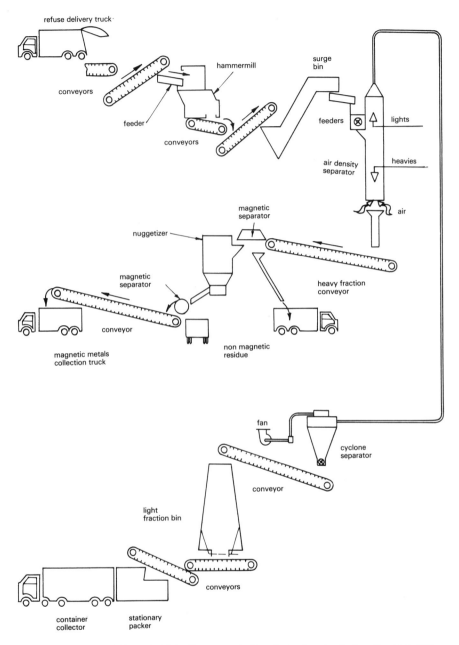

Figure 91 St Louis Missouri (USA) diagrammatic flowsheet of mechanical sorting operation

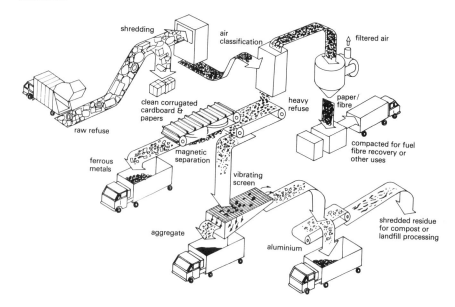

Figure 92 Resource recovery at the Monroe County

reflectance The ratio of the radiant energy reflected from a surface, to the radiant energy incident upon that surface.

reflectivity See *Emissivity*.

refractoriness Degree of resistance to softening under the action of heat offered by a refractory material.

refractory Relates to any material which is difficult to fuse, melt or soften. The primary characteristic of a refractory is the ability of the material to withstand the action of heat without deforming or softening. Different materials vary in their ability to resist softening, and the degree of the resistance offered by each material is called its 'refractoriness'.
The standard method of determining the melting point of a material (also stated as the pyrometric cone equivalent, PCE) is by comparing its behaviour under increasing temperature with that of a series of numbered standard pyramids or pyrometric cones of varying silicate composition, whose softening characteristics are known. Lauth and Vost of Sevres, France, introduced the cone system of laboratory tests in 1882, but it was not until 1886 that it was used for general work by Dr Herman A. Seger of The Royal Porcelain Factory, Berlin, in the Seger Cone system of

refractory classification. This was initially used for the determination of kiln temperature at which ceramic products were fired; the upper part of the series was also used for the determination of refractoriness of fire-clays.

refrigeration compressor Has the purpose in a vapour-compression cycle of accepting the dry low-pressure gas from the evaporator and raising its pressure to that required at the condenser. (See figure 94.)

refrigeration compressor – dynamic (centri-fugal) type The suction vapour enters the centre of the rotor and is impelled outwards by centrifugal force, leaving the tip at a relatively high speed of about 100m/s. The gain in dynamic energy is converted to the pressure differential between suction and discharge; it is a function of the tip speed and of the gas density.
Operates at low pressure, using low-pressure refrigerants and works at low compression ratios. Because the final gas velocity is high, such machines are limited to sizes of 300kw and upwards.
Capacity reduction available down to 10 per cent to 15 per cent of maximum. Compressor has a gland or an oil seal, which keeps the shaft-bearing lubricant separated from the refrigerant; it is thus almost oil free.
Application: air conditioning in large systems.

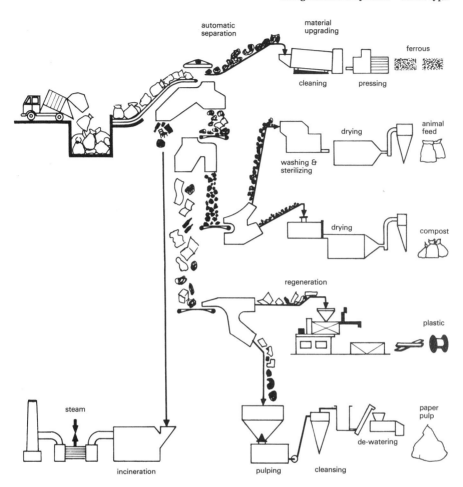

Figure 93 Rome recycling complex overall system

refrigeration compressor – hermatically sealed
Omits shaft gland, through which some leakage of refrigerant is almost certain (and unacceptable for small refrigeration systems). Has the rotor of the drive motor integral with an extended crankshaft; the stator is fitted with an extension of the crankcase.

Small-size compressors will be fully hermatic; ie all the motor and working parts are inaccessible within a sealed steel shell. Failure of the built-in motor will contaminate the system; hence internal and external motor protection devices are provided to switch off the electric supply before damage is caused.

Upon failure, on-site repair is not practicable, and complete replacement is usually necessary.

refrigeration compressor – reciprocating

Positive displacement type, relies on piston activity for operation. Modern compressor assemblies tend to be of the multicylinder type, employing six, eight or more cylinders in one packaged plant. Size of this type commonly up to 300kW.

refrigeration compressor – screw type A development of the gear pump to suit it to pumping gas. The rotor shapes are modified to provide maximum swept volume, and the helix pitch is such that the inlet and outlet ports can be located at the ends (instead of at the side). Solid portions of the helix (screw) slide over the gas ports to separate one stroke from the next; no extra inlet or outlet valves are required/fitted. Commonly twin-meshing screws are fitted on parallel shafts.

151

refrigeration compressor – sliding vane type

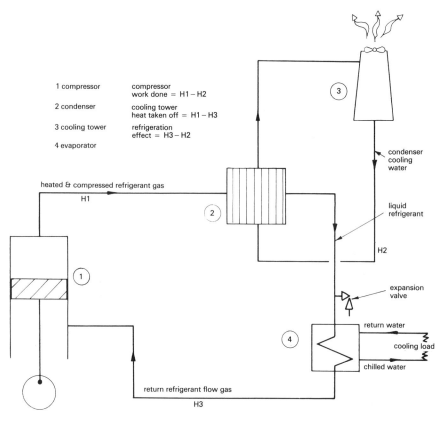

1 compressor — compressor work done $= H1 - H2$

2 condenser — cooling tower heat taken off $= H1 - H3$

3 cooling tower — refrigeration effect $= H3 - H2$

4 evaporator

heated & compressed refrigerant gas

H1

condenser cooling water

liquid refrigerant

H2

expansion valve

return water

cooling load

chilled water

return refrigerant flow gas

H3

Figure 94 Schematic of operation of the compression refrigerator

Screw compressors have no clearance volume and can work at high-compression ratios without loss of volumetric efficiency. Capacity reduction of this type permits output variation down to about 10 per cent of maximum.

refrigeration compressor – sliding vane type
Incorporates an eccentric rotary piston and sliding vanes; the latter will vary in angular alignment to provide a form of positive displacement compressor. Larger models have eight or more vanes and function without the need for inlet or outlet valves.
Limited in commercial use to an upper limit of capacity of 5kW.

refrigeration condenser Accepts the hot high-pressure gas from the compressor (in a vapour compressor system) and cools it to achieve the condensation of the gas back into liquid refrigerant. Removes superheat before latent heat

can be withdrawn. Liquid leaving the condenser may be slightly sub-cooled (below the temperature of liquification).
Cooling medium circulated through the condenser will be water (pumped) or air (fan-assisted). This may be employed for heat recovery or heat pump sink.

refrigeration condenser – evaporative
Essentially a cooling tower. See *Cooling tower.*

refrigeration evaporator Receives low-temperature, low-pressure refrigerant fluid from the outlet of the expansion valve. The liquid evaporates and takes up its latent heat from the refrigeration load (water or air); it leaves the evaporator as a dry gas on its way to the compressor of a vapour-compression system. Evaporators are classified according to function and refrigerant flow patterns. The oil must be removed from the evaporator and re-circulated to the crankcase.

refrigeration evaporator – air-cooled Similar to air-cooled condenser, using finned-tube heat exchangers and fan assistance.

refrigeration evaporator – dry expansion
Refrigerant is totally evaporated in one pass, and the oil is recirculated to the compressor at a constant gas velocity.

refrigeration evaporator – flooded Has a body of refrigerant fluid boiling in a random manner; the evaporating refrigerant leaves at the top.

refrigeration evaporator – liquid cooling
Employs commonly shell-and-tube or shell-and-coil evaporators with water as the cooling medium.

refrigeration evaporator – plate type Formed by cladding a tubular (serpentine) coil with sheet metal and welding together two embossed plates; alternatively use is made of aluminium extrusions. Extended flat plate may then be used for air or liquid cooling (the latter by immersion inside a tank).
Application: for cooling of solid articles by conduction: these are formed in rectangular packages and held close between a pair of adjacent heat-exchange plates.

refrigeration receiver Drain tank located directly after the condenser to hold reserve of refrigerant liquid with object of avoiding the backing up of excess liquid at the condenser to ensure that only liquid refrigerant enters the expansion valve.

refrigeration – ton See *Ton of refrigeration.*

refrigerator – absorption See *Absorption refrigerator.*

refuse incinerator Incinerator especially adapted for the burning of refuse and incorporates designs to overcome the difficulties which are inherent in the burning of refuse. These include:
refuse may be damp and may include up to 40 per cent water;
refuse may be variable in size, moisture content and calorific value;
refuse density varies according to its composition and origin – may be bulky;
refuse is a mixture of soft and hard constituents and is likely to contain a mix of substances from dust to large furniture sections;
refuse may include components with a low melting point, such as glassware; this could lead to rapid clogging of the fire bars and other furnace parts;
refuse may include noxious materials, such as plastics and rubber.

regeneration Relates to water softeners of the base-exchange type. The water treatment process is reversible, and exhausted exchange material can be restored to its initial condition by contact with a solution of a suitable regenerant substance.
The frequency of regeneration depends on the particular application; it is not unusual to regenerate once in every 24 hours. Regeneration may be by manual manipulation of valves or be conducted automatically under a timed programme control.

reheat *Applied to air conditioning and mechanical ventilation:* the addition of heat to the centrally treated air to meet the temperature requirement of local zone(s).
Applied to steam turbines: passage through boiler reheater section of the exhaust steam from the high-pressure cylinder of turbine with the object of reheating the steam to its original temperature before it passes into the intermediate – pressure cylinder of the turbine.

reheater (air) Heater battery which boosts the heat output (temperature) of a supply air stream between the main air handling plant and the point of supply into the environment.
Applications: recovery of design temperature after cooling in duct run; provision of higher temperature air supply to special application or area.

reinjection well Refers to geothermal schemes. That well through which the cooled water is returned to the underground aquifer.
Most geothermal schemes use a reinjection well, as the geothermal fluid is generally corrosive, highly saline and would be difficult to dispose of via the public sewers.
The reinjection well keeps the aquifer charged, so that pressure should be maintained throughout the life of the geothermal system.
The following risks must be guarded against: plugging of the surrounding strata by precipitation of dissolved solids; short-circuiting of the cooled water, resulting in premature lowering of the output temperature of the production well; energy imbalance of the scheme due to power consumption of the reinjection pumps.
It is considered that the balance of advantage lies with geothermal schemes employing reinjection wells.
Usually, the points of entry and extraction in

the aquifer are spaced about 1 km apart horizontally to avoid interference and short-circuiting between the two wells.

relative humidity See *Humidity – relative.*

relay (electric) Device which is operated electromagnetically by the electric current in *one* circuit which causes electrical contacts to open or to close with the object of controlling the flow of current in *another* circuit.

remote control and monitoring equipment A range of modules capable of data transmission, enabling remote control of heating plant, etc subject to analogue or digital measurements. *Applications:* automatic control, subject to sensors, of pumps, valves, contact breakers, ventilators, etc.

renewable air filter See *Air filter – renewable type.*

renewable energy Deposits of oil, gas, coal and uranium are finite (however one may dispute the extent of this finality) over a relatively short span of time. Renewable energy sources are (in human terms) infinitely renewable; these include wave, solar, wind, geothermal and tidal energy.

resistance (electric) Offered by an electrical conductor or appliance to the flow of current through it when an emf is applied to it.

resistance heating Space or process heating by direct electric heating through the application of voltage to the resistor element (or elements). Resultant energy conversion close to 100 per cent (there are some heat losses from the heater assembly).

resonance The build-up of excessive vibration in a resilient system. It occurs when the machine speed (disturbing frequency) coincides with the mounted machine natural frequency, or support system.

resonant frequency (Hz) The frequency at which resonance occurs in the resilient system.

return water boost pump See *Boiler return water boost pump.*

reverberation Reflected sound in a room that decays after the sound source has stopped.

reverse osmosis See *Osmosis – reverse.*

reverse return system This arrangement greatly eases the balancing of a heating or cooling piped circulation by ensuring that the length of the circuit to each terminal unit, and back, is about the same. (See figure 95.)

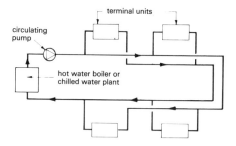

Figure 95 Reverse return system – diagrammatic arrangement

reversing valve Arranges for the change over of an operating mode; eg changeover from heating to cooling in a heat-pump system.

Reynolds number (Re) A dimensionless parameter providing a criterion for dynamic similarity in fluid flow experiments.

$$(Re) = vd/v$$

where v is the average velocity, d is the diameter (or the hydraulic radius × 4) and v is the kinematic viscosity.

rhomboidal air controller Suitable for fitting at air-discharge terminals. Ensures perfectly balanced distribution. Comprises series of vanes which are rhombic in cross-section. Adjustable from fully open to fully closed positions by regulating screw in each corner of the controller; setting each screw individually, a variety of air-flow patterns can be achieved to meet specified throw and air-distribution parameters in the ventilated space. Also used on extract ducts where the differential control feature can establish an even exhaust velocity along length of grille.

riddlings Small-size pieces of solid fuel which fall between the individual firebars into the ashpit without achieving combustion.

rigid foam thermal insulation Based on phenolic resin, non-ignitable in normal air and having low smoke emission in fire. *Applications:* wall and ceiling lining, insulated floors, core for light-weight resin concrete, cold stores, pipework and vessels for a range of industrial processes, from cryogenics to 150°C (302°F) and above.

ring circuit Electric distribution circuit arranged in the form of a ring which starts from and returns to the sub-circuit fuse or circuit breaker.

Ringleman chart Relates to the indication of smoke emission from chimneys. Ringleman chart is graduated into a fixed number of shades or smoke numbers against which the smoke is compared. The lower the smoke number, the less dark smoke in the chimney gas.

ring main Pipe system which maintains a closed loop circulation and feeds individual appliances connected to the ring main.
Applications (typical): fuel oil ring main which circulates heated oil around a boiler room to maintain the oil in a suitable flow condition. Individual burners are connected into the ring main with short connections; these may be trace-heated to maintain the correct oil temperature right up to the oil burner inlet. Boiler flow and return circuits which are pumped around the boiler room and from which separate circulations are connected and provided with separate pumps for heat supply to individual zones.
Unheated pumped oil supply pipe conveying gas oil to individual houses on a housing estate.

risers Pipes or ducts which convey water or air vertically through a building.

rock bin Form of heat storage which contains fist-sized rocks which absorb heat from solar-heated air forced through the bin under fan-power. The heat is released when the air flow is reversed.

rock-fibre insulation Rock fibre in the form of slab, pipe section, flexible rolls, mattresses and quilts.
Applications: for industrial, building and marine applications from sub 0 to 950°C (1,742°F).

roll bond process Bonding sheets of metal by simultaneous rolling and heat-treatment process.

roll-type air filter See *Air filter – roll-type.*

roof pond Offers a passive form of heating and cooling. Comprises bags filled with water which provide the required or desired thermal storage mass on the roof.

roof-top boiler plant A recent development in which the boiler plant room location is moved from inside (or below) the building to the roof;

commonly, this then forms a plant complex which is visually integrated with the lift motor room and the cold-water storage.
Advantages: good access generally possible around the plant items; ease of ventilation (without recourse to mechanical ventilation); plant does not occupy valuable building space.
Disadvantages: roof slab must be strengthened to support plant weight; floor below plant must be constructed to avoid transmission of vibration into building; plant acoustic levels must be specified and controlled; precautions must be taken to ensure that the boilers cannot be drained accidentally; thermosyphonic heat circulation is not possible; safe all-weather access is required to plant room.
Where adequate space is available on a roof, the advantages often outweigh the disadvantages; the latter can be dealt with by appropriate engineering techniques.

room thermostat Accurately controls space-heating temperatures. An accelerator heater can give minimum space temperature variation.
Applications: models are available for the control of pumps, relays, gas and oil burners, motorised valves, air-conditioning equipment and other devices requiring on-off or change-over switching.

Rotameter Measuring instrument by Rotameter Ltd which establishes rate of flow of a fluid at any time. Essentially comprises a specially shaped float resting on the fluid inside a tapered vertical glass tube. Float may incorporate spiral grooves in its upper periphery which keep the float in rotation about a vertical axis centrally inside the tube as the fluid being measured flows past. The glass tube has an engraved velocity or quantity scale and the position of the float relative to this scale provides an immediate direct indication of the mean rate of flow. No calculations are required, except to apply a correction factor to allow for non-standard conditions.
Application: for measurement of both liquid and gas flows. Made in various sizes suitable for measuring gas flows between 0.006m³/hour to 85m³/hour, the larger size can be inserted in pipelines of 200mm diameter. Larger sizes have the vertical tube made of metal and the float then incorporates a vertical sighting rod which is visible together with the scale above or below the instrument in a glass-fronted sighting compartment. Widely used in the measurement systems of process lines. Serves particularly useful purpose as an indicator in a system in which the flow has to be maintained at a closely monitored specified rate.

rotary burner Oil burner operating with low pressure air; the oil is distributed over the air blast from the lip of a rapidly rotating (or spinning) cup type fitting. Can handle very large oil inputs.

rotary valve Provides a suitable method of control over the quantity and rate of the waste fuel (wood chips, sawdust, etc) fed from a silo (storage hopper) into a furnace. The speed of rotation of the valve controls the flow of the material via a direct-current electric motor with non-overloading infinitely variable speed facilities. (See figure 96.)
The blades of a rotary valve are usually fabric-ated either from heavy-gauge steel plate with heavy-duty rubber (or neoprene) tips or from wholly fire-resistant heavy-duty rubber or neoprene to achieve safe and positive separ-ation between the stored waste (which is at low pressure) and the high-pressure furnace feed line, to provide a fire-break.

run-around heat recovery coils Used on a run-around basis, with one coil in the exhaust air ductwork transferring a proportion of the heat to circulated water or water/glycol solution, which is pumped through another coil in the supply air ductwork.
Applications: in those applications requiring a high proportion of introduced fresh air and positive removal of exhaust air by mechanical means.
Energy-saving potential: typical efficiencies of 50 per cent to 60 per cent are possible.

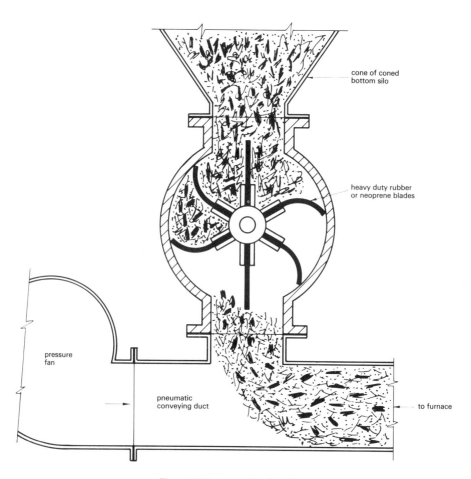

Figure 96 Rotary valve function

sacrificial anode Large lump of steel or other appropriate metal connected to a metal pipe-line to serve as anode in a cathodic corrosion protection system. Corrosion subsequently occurs at the sacrificial anode and not on the pipe. In a very corrosive environment, such anodes are buried in the ground, using a surround of coke breeze as backfill.

salinity Salt (saline) content of a liquid; usually given in terms of g/litre. Of particular relevance to geothermal energy schemes.

sander dust Finely powdered wood-waste collected off finishing or sanding machines. It is highly explosive, being spontaneously ignitible at about 600°C (1,112°F). It is usually collected in a separate dust-collection system and cannot be safely incinerated in common types of incinerators without risk of dangerous explosion.

Sankey diagram Provides a graphic statement of a heat balance for a process or boiler plant. (See figure 97.)

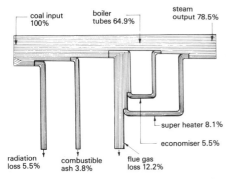

Figure 97 Typical Sankey diagram

saturated steam See *Steam – saturated*.

screen (electric) A shield which isolates electrical equipment, circuits, etc from the undesirable effects which could be caused by external electric or magnetic fields. Similarly applies to screened cable.

sealed heating system Relates to wet heating and hot-water installations; refers to a piped system which is not permanently vented to atmosphere and does not incorporate an open feed and expansion cistern to automatically replenish water lost from the system.
Such a system incorporates a pressure vessel, appropriately sized for the water capacity and pressure of the particular installation. The pressure vessel has a diaphragm which separates its water space from the upper space which contains a pressurised air or nitrogen cushion. Water expands against the diaphragm during heating up.
Commercial size sealed systems operate with a water break-tank and automatic controls which switch a boost pump to compensate for loss of pressure due to leaks at valves and pump glands. Domestic size systems usually are provided with a flexible connection which is *temporarily* connected into the mains water supply to boost pressure.
The satisfactory operation of such systems requires the provision of pressure gauges in accessible locations, so that the pressure may be closely monitored. Safety valves are required.
Sealed systems must be used with micro-bore pipe systems to exclude all particles of corrosion products. Sealed systems in general offer the advantages of exclusion of contact between the water and air and obviate the need for open vent pipes (particularly where such vent pipes are difficult or costly to accommodate). However, they present the system user with an added complication.

Caution: the correct selection and specification of the pressure and water capacity of the pressure vessel and the provision of pressure gauges are essential for the safe operation of the sealed system.

seals – mechanical Used with circulating pumps, replacing conventional pump glands. Do not leak water and do not require pump drains. Failure of seal may cause major water leak at pump.

S

secondary air As applied to a fuel bed, is introduced *above* the bed of solid fuel to ensure the complete combustion of thick bed fuel layers and (more importantly) of fine fuel particles in suspension above the bed. With certain types of fine-particle fuels (such as sawdust and sander dust), the secondary air supply requires special consideration because as much as 75 per cent of the fuel feed may have to be burnt in suspension above the fuel bed.
In liquid- or gas-fuel burning, the secondary air is introduced *around* the fuel-primary air mixture to ensure complete fuel combustion and also to assist in the establishment of the flame form or pattern.

secondary circuit Separated from the main energy source by an intermediate heat exchanger.

Application: draw-off domestic hot-water system fed off indirect hot-water cylinders.

second law of thermodynamics It is contrary to all experience to find a self-acting machine able continuously to transfer heat from a cold body to another body at a higher temperature.

sectional boiler Cast-iron or mild steel assembly of individual matched sections to form one joint boiler unit. Sections are joined at the water-ways with nipples. The combustion interfaces are closely machined to make a tight fit; additionally, the joints are secured with boiler compound, asbestos rope, etc.
Sectional boilers offer the particular advantage of permitting installation in boiler rooms with restricted access, as each section can be handled individually into the boiler room for assembly and testing *in situ.*
Sectional boilers are made in maximum sizes to capacities of about 1,000 kw. Site assembly is relatively inefficient and costly (compared with factory assembled and tested boilers).
Suitable for steam generation in closed system only. (See figure 98.)

selective absorber A dark surface having a high absorption coefficient for short-wave solar radiation, but a low emission coefficient for long-wave low-temperature radiation.

selective surface A surface which has high absorbance for incident solar radiation (wave-lengths less than 1,000nm) and high reflectance (low absorbance) in the infra-red range (wave-lengths greater than 1,000nm).

semi-silica A refractory of limited application. Made from fire-clay and a large proportion of silica grog. It has a low melting point and poor resistance to slag attack. Its resistance to spalling is good, and it is resistant to load at moderate furnace temperatures. The usual tendency of the clay to shrink is offset by the corresponding tendency of the silica to expand to tridymite. The result is a material which shows a constancy of volume in service and is suitable for use in furnaces where the temp-erature is not too high and where there can be no slag attack.

sensible heat See *Heat – sensible.*

sequence control for boilers See *Boiler sequence control.*

sequential metering Venturi principle based flow meters do not provide acceptable accuracies when operating below 20 per cent of rated meter capacity. To achieve greater accuracy, sequential metering is adopted. With this, several flow meters are installed in parallel and are arranged to turn the appropriate meter(s) automatically on and off in a sequence determined by the nature of the varying steam load. The control may be either electrical or by steam pressure.
First cost of such electrical metering system is

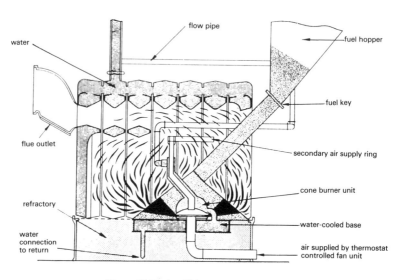

Figure 98 Earlymill burner arrangement

high; it is therefore best suited to the larger systems in excess of 10 tonnes of steam per hour.
Steam pressure control is less costly and therefore better suited to the smaller systems.

serving (electric) Relates to electric cables. Layers of fibrous material, permeated with waterproof compound, applied to the exterior of an armoured or metal-sheathed cable for additional protection.

servo-element Relates to automatic control applications. Acting on the control impulse, it resets the controlled element (eg valve, damper, etc) to return the controlled function to the control set point.

servomechanism Amplifies small signal impulse from a measuring or detecting instrument and translates same into the greater force required to effect the consequent control by operating a damper, valve motor or other equipment.

settlement chamber Fitted to boilers and furnace systems before entry of flue gases into the chimney. Consists of an enlargement of the flue ducting or a separate chamber in the flue gas path. Reduces the gas velocity and thereby allows the larger grit particles to settle out and be subsequently removed.

sewage gas Generated during the natrual process of digestion in sludge tanks at sewage farms, in the form of methane. Collected and used as fuel. See also *Biogas.*

shading Means of reducing solar-heat transmission into a space. Utilises internal, external or shades (blinds); also can be located between double glazing panes. Essential on southern building face to obviate glare and to reduce costs of air conditioning.

shading coefficient Relates to shading provided by different fenestration arrangements; it is usually taken to be the proportional amount of the maximum heat gain through the fenestration other than 3mm thick single clear glass. Varies from 1.0 for 3mm clear glass to about 0.2 for double glazing with reflective coated glass and interior shading.

shallow bed Bed depth of less than 456mm can be considered as shallow bed. Deeper bed is referred to as deep bed system.

shell boiler Such as the Lancashire or economic boiler, comprises a shell in which the

water being heated is stored and a set of flue tubes which traverse the shell and through which the flue gases pass on their way to the chimney. Convection heat transfer is from the flues to the stored water. The limit of working pressure for such boilers is about 20 bar and of output 9,000kg of steam per hour in a single boiler.

sheradizing Rust-proofing process using low-temperature heat treatment with zinc dust to produce a hard, uniform coating of iron-zinc alloy suitable for the anti-corrosion protection of a wide range of ferrous components.

short-circuit (electric) Connection made accidentally (usually) or otherwise between two points in an electrical installation which have a difference of electric potential between them, the points of the short-circuit connection being of sufficiently low resistance to allow a very much larger current than normal to flow through the circuit established by this connection. Fuses and other protective devices are fitted to protect circuits and sub-circuits from the possibly harmful effects of short-circuits. *Applied to flow of fluids (water, air, etc) within pipe systems:* incorrect circuit connection or balancing which causes the fluid to bypass sub-circuit(s).

shunt connection 1. As used in electrics, the connection of two electrical circuits in which the same emf is applied to both.
2. As used in mechanical enginering, a bypass pipe or duct.

shuntflo steam meter With this type of meter, a shunt path (bypass) around the steam meter passes a fraction of the total steam flow, driving a vaned wheel at a speed which is proportional to the total steam flow through the meter; the vaned wheel is geared to indicate the total volume of steam which has passed through the meter. The top cover, which embodies the shunt path arrangement around the meter orifice, is identical for all meter sizes; hence, the ratio of flow through the meter and through the shunt varies with different sizes of meters. The actual steam flow rate through the meter is determined from a series of basic equations.
This type of meter is suitable for large rates of steam flow, as only a small proportion of the steam flows through the shunt bypass.

sight glass Fitting with transparent observation port.
Application: in condense systems placed after the steam trap to permit a view of its function.

silencer See *Attenuator*.

silica (From Latin 'silex', meaning 'flint'.) Constitutes about 12 per cent of the minerals which form the rocks of the earth crust. It can exist in many forms, the most common (related to fire-bricks) is quartz, the modification of silica which may be formed at relatively low temperatures, below 820°C (1,598°F).
The sensitivity of silica to abrupt changes in temperature restricts its use as a refractory. It contains glass of high melting point and retains its stiffness under load to within a few degrees of the melting point and is, therefore, suitable for high furnace temperatures. It has a strong spalling tendency at low temperatures below dull redness.

silica gel Dehydrating agent; generic term which covers many forms of the material which can be obtained by acidifying sodium silicate solutions. (See figure 99.)
The material changes colour when saturated. It can be reactivated by the application of heat which drives off the moisture.
Application: air drying; it is usual to have two silica gel beds in parallel – one operational, the other reactivating.

silicon carbide A refractory (SiC) used extensively in high-temperature and abrasive applications, such as in cyclone wood waste incineration plants. It is a carborundum (proprietary name) refractory made in an electric furnace, sorted, crushed, mixed with a clay binder, shaped and fired. It does not fuse even at temperature as high as 2,700°C (4,892°F), but it decomposes slowly at temperatures in excess of 2,200°C (3,992°F).
Silicon carbide is very strong, hard and resistant to abrasion; its coefficient of expansion is low and uniform; it has good resistance to spalling and to the action of most slags.
High cost restricts the application of silicon carbide; it is usually employed in a cyclone furnace only as the inner skin, the cheaper alumina bricks being used as the intermediate and outer linings.

silo Storage hopper located above ground and commonly constructed of steel or cast-iron

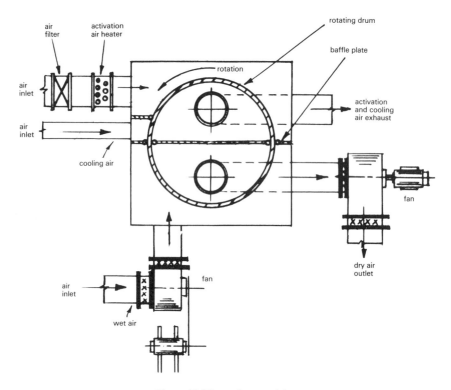

Figure 99 Silica-gel rotary drier

sections. Two types of silos are available for storage of waste, fuel, etc: cone-bottomed and flat-bottomed.

Cone-bottomed silos are designed to avoid the bridging within of the waste material and to allow a free flow of this to the bottom outlet, usually via a rotary valve.

Flat-bottomed silos depend on the positive displacement of the waste by means of a

rotating worm or screw-feeder which is fitted inside the silo. They offer the obvious advantage of providing greater storage for a silo of the same diameter. (See figure 100.)

single-phase circuit Applies where a single alternating current is supplied by one pair of wires (conductors). Generally used with domestic and small industrial or commercial systems.

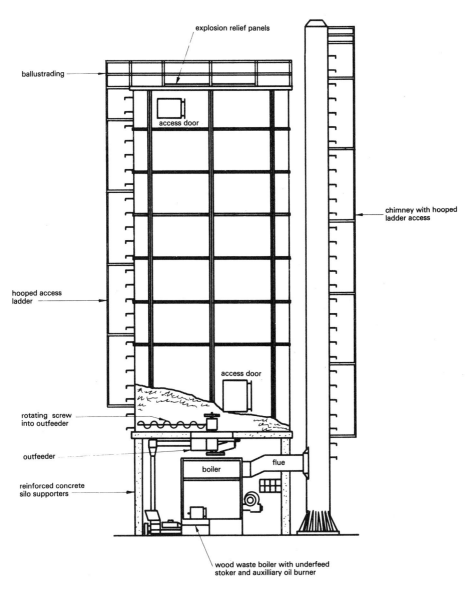

Figure 100 Flat bottomed silo

161

single-pipe district heating Operates without the recirculation of the water heating medium. The hot water from the boiler plant is pumped first through the domestic hot-water heat exchangers, then through the space heaters and from there to waste after all possible heat abstraction has taken place. Requires effective water treatment to protect the heat-exchange surfaces from hard scale due to the continuing change of water in the system. System offers considerable economies in first cost and in operation; depends on availability of cheap water supply and on consent for water wastage from the water utility company. Used extensively in Soviet Union in conjunction with magnetic water treatment.

single-pole switch Circuit breaker, cut-out, fuse-switch and similar in which the circuit is broken in one pole only.

S.I. (units) S.I. stands for Système International d'Unités. (See B.S. 3736:1964)

The basic S.I. units are m, kg, degree Kelvin, ampere, candela, radian and steradian for angles (plane and solid). All other units are derived from these.

Prefixes are necessary for very large and very small units, thus:

T(era) (10^{12})
G(iga) (10^{9})
M(ega) (10^{6})
K(ilo) (10^{3})
h(ecto) (10^{2})
c(enti) (10^{-2})
d(eca) (10^{1})
d(eci) (10^{-1})
m(illi) (10^{-3})
(micro) (10^{-6})
n(ano) (10^{-9})
p(ico) (10^{-12})

Associated terminology uses the following units:

N – 1 Newton = 1 kg m/sec^2 (force)
J – 1 Joule = 1 Newton m (energy)
W – 1 Watt = 1 Joule/sec (power)
K – Degrees Kelvin (Centigrade in degrees absolute) (temperature)
H_z – 1 Herz = 1 cycle/sec (frequency)
C – 1 Coulomb = 1 ampere/sec (electric current flow)

skin effect See *Induction heating*.

skirting heater See *Baseboard heater*.

slaked-lime injection flue gas cleaner Simplified method of cleaning flue gases with reduced energy consumption. Process indicated on figure 100A. Plant (by Sakab, Sweden) is a combination of spray absorber and electrostatic precipitator. The flue gases pass through the absorber into which slaked lime is continuously injected. Each one litre of lime is there atomised to about 20 billion droplets of 60m^2 surface area and react with the pollutants in the flue gas (HC1, HF, SO_2 and others). The residual heat in the ingoing gases is used to evaporate the water and the reacted components thereby form a free-flowing powder which settles out in the electrostatic filter.

Applications: industrial furnaces, waste incinerators, etc.

Energy Saving Potential: due to the elimination of wet scrubber waste disposal and re-heat of the exit gases; this would arise with a conventional flue gas cleaning system.

sleeve or pipe sleeve Usually a short piece of pipe, cut and oversized to permit the associated pipeline to expand and contract without rubbing. The sleeve is built into the wall, ceiling or floor through which the pipeline passes. It is usual to have one sleeve per pipe. The ends of the sleeve project slightly beyond the confines of the structure and may be closed off with special sleeve caps to obviate entry of dirt and/or vermin.

Sleeves are commonly of metal (mild steel or copper); may also be of plastic where appropriate.

sliding expansion joint Comprises two cylinders sliding inside a pipe gland to permit pipe movement. Operates in conjunction with pipe anchors and pipe guides. (See figure 101.)

slow sand filters In these, the water is passed very slowly downwards through sand beds of about 75cm thickness. After about a month in operation, the surface has to be skimmed, but this period can vary widely depending on the condition of the raw water. Eventually the filter has to be taken out of service and refilled with washed sand.

small-bore heating system Applies mainly to pump-assisted domestic size installations employing 12.5mm and 19mm bore copper or thin-bore m.s. pipe distribution. Collecting pipes at boiler may be of larger bore.

Hot water supply storage can be incorporated to operate on thermosyphonic (gravity) circulation – minimum primary circuit pipe size: 25mm bore.

Usually employs canned rotor type circulator to meet the head requirement. Each subcircuit must be valved to permit fine balancing of water (heat) flow.

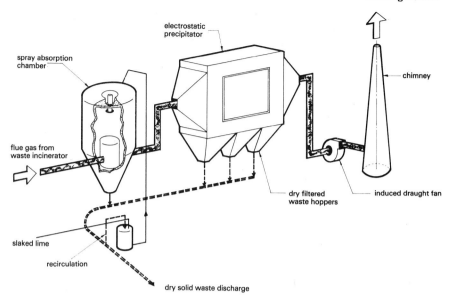

smoke generator

electrostatic
precipitator

spray absorption
chamber

chimney

flue gas from
waste incinerator

dry filtered
waste hoppers

induced draught fan

slaked lime

recirculation

dry solid waste discharge

Figure 100A Slaked lime flue gas cleaner

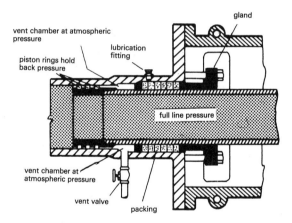

gland

vent chamber at atmospheric
pressure

lubrication
fitting

piston rings hold
back pressure

full line pressure

vent chamber at
atmospheric pressure

vent valve

packing

Figure 101 Piston ring expansion joint unit

smoke density equipment Measures smoke density in terms of light obscuration emitted from industrial boilers and furnaces.
Applications: applied to industrial boilers and furances to measure smoke emitted to atmosphere, which is also a measure of carbon or combustible gas which is being emitted to the atmosphere. (Smoke emission from a boiler indicates inefficient operation and, usually, more air being needed to promote complete combustion and burn the fuel efficiently.) (See figure 102.)

smoke generator Device for generating smoke for purposes of flow visualisation.
Application: checking air movement across spaces, eg to locate and follow draught movements; indicating air flow patterns at air supply and extract grilles, slots, etc; establishing air movement at windows, air curtains; monitoring fume and dust extraction, etc.
At its simplest, generator may consist of a smoke tube which is fitted into an aspirator; when both ends of the tube have been broken

smut emission

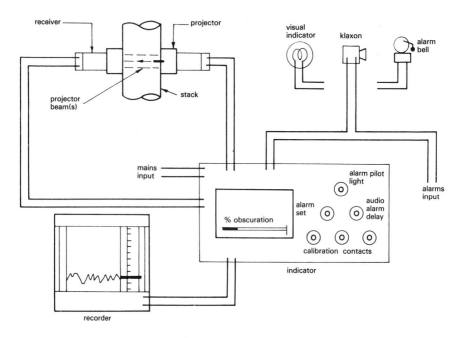

Figure 102 Smoke density measurement

off, the smoke is aspirated. Used for estimating the velocity of slow-moving ventilating currents by timing the travel of the smoke given off by the tube when air is aspirated through it; can also be used to indicate the direction of air currents.

For research and the more sophisticated smoke testing, electric mains operated smoke generators are available, such as the TEM Smoke Generator System developed by the National Physical Laboratory, Teddington, England. With this, the smoke probe operates in conjunction with an electric generator. The smoke is generated at the extreme tip of the probe by the vapourisation of a special oil. Smoke is then emitted directly into the airstream from a compact screw-on vapouriser that is fed with oil and low voltage power via the thin stem of the probe. The standard probe is 368mm long with a wake-minimising curve into alignment along the airstream. Equipment is available for producing clearly visible smoke plumes at flow rates of up to 90m/s. *Applications:* for use in wind tunnels, industrial and architectural aerodynamics, air conditioning, fume and dust extraction, heat exchanging, etc.

smut emission From chimneys into the surrounding atmosphere causes pollution through the eventual deposition of unburnt or partially burnt combustion products. Caused through inefficient combustion (which passes unburnt particles into the chimney) and through acid condensation brought about by cooling in the flue gas (or chimney) system below the acid dew point of the gas. Can be avoided by good standard of combustion management and by suitable thermal insulation of the chimney system.

sol-air temperature The external temperature of a hypothetical uniform environment which has the surroundings and the air at equal temperatures and which would provide the same rate of heat transfer through a building element as occurs under the actual prevailing conditions.

solar array A number of individual solar-collection devices arranged in a specific pattern to form a collector complex.

solar chimney Operates on the principle of guided convection. Air is drawn up through the solar chimney, activating one or more turbines for the generation of electricity in its passage through the chimney.

The essential components of a solar chimney are: a central chimney, a greenhouse formed of transparent plastic materials which is supported a few metres above ground by a metal frame and which surrounds the chimney close to its base.

In operation, solar transmission through the plastic screen causes the air which is trapped in the greenhouse to heat up so that a convection system is established in which this hot air is drawn upwards through the central chimney at a velocity of 20 to 60 m/s turning one or more turbines. The heated air is continuously replenished by cooler fresh air drawn in at the periphery of the greenhouse.

A pilot plant of output capacity of 100 Kw has been established in Manzanares in Spain (midway between Mardrid and Almeria) and is now operational. The chimney is 200m high and of 10.3m constant diameter. The area of the solar collector (greenhouse) extends to a radius of 126m from the chimney and varies in height from 2m at the circumferance to 6m at the centre.

The life of the plastic sheeting is critical to the success of the system, as it represents about 45 per cent of the total investment. Most rigorous selection and evaluation of the plastic material is therefore essential.

On test, the pilot plant has performed well. Plastic materials being tested include polyvinyl chloride and fluoride and Tedlar. A drainage valve has been fitted at the centre of each 6m² sheet – this is normally closed to prevent the loss of air, but during periods of rain it opens and permits drainage and cleaning of the upper surfaces of the collector of dust and other accumulated foreign matter; it is thus virtually self-cleaning (this feature represents a useful maintenance advantage over solar power systems which rely on mirrors, solar collector surfaces, etc).

Professor J Schlaig of Stuttgart (Germany), whose papers have initiated the project, estimates that the present-day costs for larger installations will be about £(sterling)230/Kw and with associated operating costs of £0.01 to £0.02/Kw/hr. It has been established that the output capacity of the solar chimney plant increases (and the costs decrease) with increasing temperature below the collector (greenhouse) roof and with increasing tower and collector dimensions.

A plant of 1,000MW is considered to be feasible with a chimney height of 900m; its collector would have a diameter of some 10km. The required plant area would then be 80 km²; the ideal size for the first generation of solar chimney commercial plants is likely to be 270Mw with a chimney height of 800m. It is considered that the ideal locations for such plants will be rock deserts where the solar radiation is between 500 and 600 w/m², allowing almost unlimited power generation from a solar chimney system. This type of plant could also be viable in Southern Europe and in North Africa where the solar radiation is at the lower level of 400 w/m² and there are above 300 days of sunshine per annum.

solar collector – concentration ratio The ratio of the heat flux within the image, to the actual heat flux received on earth, at normal incidence.

solar collector – concentrator A reflector system to increase sunlight intensity on a given area. (See figure 103.)

solar collector – efficiency An indication of relative performance or character, usually indicated on a graph incorporating important variables shown in the Hottel-Whillier-Bliss equation:

$$Q = F(aI - UAT)$$

in which

- Q = useful heat collected
- F = heat transfer effectiveness
- a = transmittance absorptance product
- I = incident solar radiation
- U = heat loss coefficient
- AT = temperature difference (collector mean temperature – ambient air temperature).

Collector efficiency varies from sunrise to sunset and especially when the collector is of the fixed, non-tracking type. Normal increase of Delta T also seriously affects the performance of some collector types.

solar collector – evacuated tube An alternative approach to reducing collector heat losses by employing a partial vacuum within transparent tubes arranged in parallel form over a reflector plate. Within each tube are usually one or more absorber tubes.

solar collector – flat plate Any non-focusing, flat-surfaced solar heat collecting device. (See figure 104.)

solar collector – parabolic A focusing type of solar collector, usually arranged in trough form, having a line-focus.

solar collector – paraboloid A focusing type of solar collector produced from the rotation of a parabola around its axis. The concentration rate is the square of that for the parabola. (See figure 105.)

solar collector

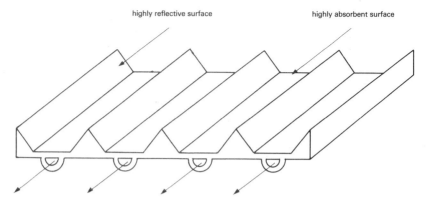

highly reflective surface   highly absorbent surface

Moderately concentrating flat plate collector

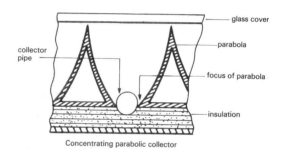

glass cover

collector pipe

parabola

focus of parabola

insulation

Concentrating parabolic collector

reflecting spiral curve

collector tube

Moderately concentrating trapezoidal collector

Figure 103 Concentrating collectors

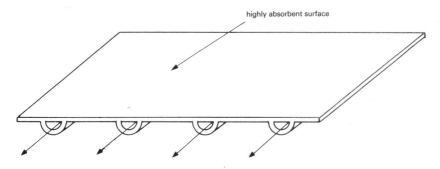

highly absorbent surface

Figure 104 Conventional flat plate collector

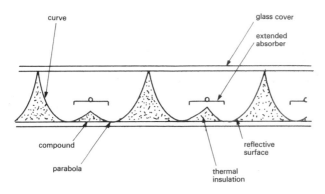

curve

glass cover

extended absorber

compound

parabola

reflective surface

thermal insulation

Concentrating collector assembly

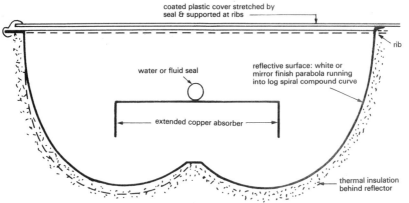

coated plastic cover stretched by seal & supported at ribs

rib

water or fluid seal

reflective surface: white or mirror finish parabola running into log spiral compound curve

extended copper absorber

thermal insulation behind reflector

above illustration is subject to various patent applications

Note because of certain scattering of irradiance through plastic approximated curve, white reflector is sufficient. If mirror finish reflector is used then curve must be mathematically correct. Glass is a better cover than plastic.

Figure 105 Parabolic focusing collector general principles

solar collector – tracking A mechanised solar collector arranged to follow or track the path of the sun and normalise the angle of incident radiation falling upon the collector surface.

solar constant The intensity of solar radiation outside the earth's atmosphere at the mean distance between the earth and the sun; equals 1.353 kW/m^2 or 1.353 $\times$ 10^3 J/sec/m^2 or/ 1.940/min/cm^2 or 428 Btu/hr/ft^2.

solar energy – active system See *Active system.*

solarimeter See *Pyranometer.*

solar pond An artificially formed pond designed specifically to attract and trap solar energy for subsequent energy use. Typical pond has a depth of 1m to 2m; the bottom of the pond would be painted black. A salt solution is introduced to establish a density gradient from top to bottom. The bottom zone provides heat storage and extraction.
Operating extraction temperatures of 90°C (194°F) with the surface at 30°C (86°F) have been achieved. (See figure 106.)
Can function in association with a low-pressure steam turbine to generate electricity; suitable for heat recovery or water heating or to perform both functions.
Application: in areas with ample sunshine, eg California, India, Israel (where there are solar pond installations at present).

solar screens systems Various types and arrangements available to reduce solar transmission through glazing. When of modular construction, the system comprises a number of basic units assembled to form banks of horizontal, vertical or inclined (commonly aluminium) louvres with support clips of high quality engineering plastic to provide the required supporting strength and to promote silent operation.
Louvres may be automated to follow the sun cycle and incidence, they may be manually adjustable or be fixed in the optimum direction for solar energy exclusion.
The louvres will be located externally to the building and thus deflect the solar energy before it reaches the building. Such duty imposes severe operating and weather conditions on all the components: the selection of proven quality materials and workmanship essential to the longer term success of such systems.

solar window films A variety of films are available to provide solar protection and heat insulation for all glazed areas. Ranges include solar control film and heat insulation films which combine solar heat insulation in summer and reduce heat loss in winter.
Energy-saving potential: Solar films are stated to provide up to 78 per cent reduction of solar heat and glare, with consequent saving on air conditioning. Winter heat insulation films save up to 40 per cent of potential radiant heat losses.

solenoid Wound coil arranged for producing magnetic field; used in conjunction with solenoid valves to control the flow of fluids.

solenoid valve Control or regulating valve which is fitted with an electric operating motor. May be controlled manually or automatically (by thermostat, pressurestat, timer, etc).

solenoid valve – servo-assisted Used for applications which involve high flow rates and high operating pressures.
Available with slow-closing valve seat which reduces (obviates) likelihood of waterhammer.
Applications: water-treatment plant, irrigation systems, vehicle wash equipment.

solid state (bottoming) 1. Vibration isolation, the unwanted situation when a spring can be compressed no further and the coils are in contact.
2. Electronics, the description of circuitry using semi-conductors instead of valves; eg transistors.

solstice The time at which the sun reaches its greatest declination, north or south.

soot blowing Removes fire-side deposits from the boiler heating surfaces. Used with the larger capacity water-tube boilers which operate with high furnace temperatures causing the combustion deposits to bake hard onto the surfaces.
The presence of soot deposits on the furnace and superheat tubes of water-tube boilers cannot be tolerated, as these reduce the heat transfer between hot gases and water and may cause local overheating leading to the fracture of the deposit-coated tube(s) and expensive down-time and remedial works.
The soot blowing nozzles must be carefully positioned to blow into the spaces between the tubes forming nests of tubes, avoiding direct impingement of the steam or air blast onto the thin water-tube walls of the boiler.
In large boilers, the soot blowing facility is considered integrally with the design of the boiler, as correct soot blowing is critical to the boiler efficiency and reliability. A steam

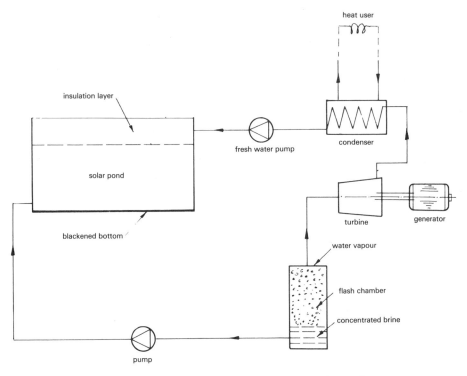

Figure 106 Extraction of heat and power from a solar pond

pressure of 13 to 35 bar is required for effective soot blowing.

The process may be manual (for smaller size boilers) or automatic (invariably so for large capacity boiler units).

Excessive soot blowing is wasteful of energy. The fire-side condition of the tubes must therefore be monitored to maintain optimum soot blowing effectiveness without wasting energy.

soot deposits Associated with incomplete combustion of fuels in boilers. Such deposits become damp when the boiler is shut down and attack the metal construction of the boiler, causing severe wasting of boiler components. Chemical analysis of soot deposits associated with wasting corrosion indicates presence of metallic sulphates and oxides, loosely bound together with carbonaceaous matter and water, as well as with some free sulphuric acid (up to about 2 per cent). Good boiler design should make all potential soot-bearing surfaces easily accessible to cleaning in a routine manner.

The presence of soot deposits on heat-transfer surfaces greatly impairs efficiency and wastes heat input energy.

spalling Term used with refractories; relates to the splintering or cracking of refractories where fragments of the brick are separated and fresh surfaces are exposed.

specific gravity The *ratio* of the weight of a substance to that of an equal volume of water at a specified temperature (usually taken at 3.9°C (39°F)).

specific heat The quantity of heat required to raise the temperature of unit mass of a substance through 1 degree temperature. Experimental evidence shows that equal masses of different substances require unequal quantities of heat to raise their temperature through the same range.

The number which expresses the specific heat of a substance is the same irrespective of the system of units which is employed. The specific heat of most substances varies somewhat over a range, depending on the temperature. For strict accuracy, a particular specific heat should be referred to the temperature at which it applies. At a temperature of 100°C (212°F), the following are typical values of specific heat for

specific heat loss rate

common substances:

aluminium:	0.219
copper:	.0936
iron:	.119
lead:	.0305
tin:	.0552
zinc:	.093
ice ($-20°C$ to $-1°C$)	.502
water:	1.00

specific heat loss rate An index of the thermal properties of a building, given as the total heat loss rate per unit temperature difference (W/°C).

spectral distribution An energy curve or graph which shows the variation of spectral irradiance with wavelengths.

spiral ducting See *Ducting – spiral*.

spiral heat exchanger Comprises a spiral element with flow channels alternately open at one side and welded at the other to ensure isolation of fluids. It is sealed between two gasketed cover plates. Hot fluid enters at the centre and flows spirally to the periphery, while cold fluid moves countercurrently.

splitters Metal vanes inserted into air ducts to direct the air flow in an even manner.

spontaneous combustion See *Combustion – spontaneous*.

spot-cooler Term commonly applied to portable packaged air conditioner of up to 3.5 kw capacity. The stream of cool air is directed at a 'spot'; this may be a receptionist sitting inside a sundrenched shop window, a worker on a production line, etc. The heat dissipated at the condenser is blown out to the rear.
Such units available mounted on movable pillars or castors; also separated into two components: the cooling section and the air cooled condenser (heat dissipator) connected together by flexible refrigerant piping.
Must be moved with great care to avoid damage to the flexibles.

spray recuperator Stainless-steel recuperator condenses the water vapour of exhaust gases by the direct spraying of water. The process reclaims sensible heat (by lowering the exhaust gas temperature) and latent heat with the condensate. (See figure 107.)
Applications: for use in the exhaust stream of boilers and in any non-acidic exhaust from dryers, etc.
Energy-saving potential: recovers 60 per cent to 70 per cent of heat normally lost to the atmosphere.

spring pipe support Vertical support bracket which incorporates a spring to absorb vibration movement.
Application: on pipework inside a boiler room or ventilation plant room to obviate the transmission of plant vibration (eg from fans or pumps) into the general pipe or duct distribution.

spring rate The force which results from the resistance of expansion bellows to movement (compression or extension) given in kg or N per unit length (mm or cm) of movement. The *spring loads* are subtracted from the main anchor loads due to pressure (under cold conditions) and added to the anchor loads due to the pressure load (under hot conditions).

sprinkler stoker Automatic mechanical stoker which sprinkles the coal onto the grate to achieve a thin and uniform fuel bed.
Mechanism comprises essentially a rotary coal sprinkler or a spring-loaded shovel.
Type well suited to the burning of coal with a high colatile content. Can handle effects of caking properties.

spur Branch cable connected to a ring circuit.

squeeze valve See *Valve – squeeze*.

standing charges Relate to utility tariffs; those components of utility tariffs which apply whether or not there has been consumption.

star-delta starter See *Electric motor starter – star-delta*.

static deflection The distance vibration isolators compress when loaded.

static regain A method of sizing ventilation/air-conditioning ducts which provides for the same static pressure at each junction in the main run.

stator That component of an electric motor or alternator which has windings and a core and does not revolve.

stator-rotor starter See *Electric motor starter – stator-rotor*.

steam Water vapour generated by evaporation of water. Term generally refers to generation of steam in a closed vessel under controlled conditions of pressure.

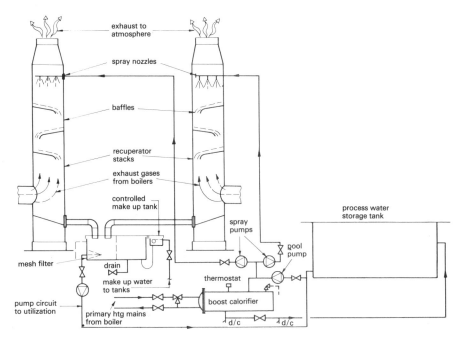

exhaust to
atmosphere

spray nozzles

baffles

recuperator
stacks

exhaust gases
from boilers

controlled
make up tank

mesh filter

drain

make up water
to tanks

pump circuit
to utilization

primary htg mains
from boiler

thermostat

boost calorifier

d/c

d/c

spray
pumps

pool
pump

process water
storage tank

Figure 107 Multi-recuperator system

steam – de-superheating When used for heating, the superheat must be removed from superheated steam; a de-superheater operates by the injection into the fitment of a spray of distilled water. Some of the spray water is evaporated in the process and increases the weight of saturated steam, leaving the de-superheater into the pipe system.

steam – dryness Relates to steam systems; the fraction of the steam which is *dry,* eg 1 kg of steam which has a dryness fraction of 'q' consists of 'q × kg' of dry steam and '(1 – q) kg' of water in suspension.

steam – flash vessel Pressure vessel into which high-temperature condensate or waste hot water is introduced and allowed to flash off into steam at a pressure associated with the input water temperature. The steam is piped off to supply equipment suitable for operation with the available flash steam pressure.

steam – from and at Term used in specifying boiler output capacity. Implies that the steam is generated *from* feed water *at* the specified temperature, eg from and at 130°C (266°F).

steam – generator (other than boiler) A pressure vessel into which high-pressure hot water is introduced via a heat transfer coil or battery and applies heat to a body of water to generate steam at some specified temperature. *Application:* generation of steam for humidification, cooking equipment and similar, which are associated with high-pressure hot-water systems.

steam – pass-out (re CHP) Exhaust steam discharged from a steam engine or turbine into a process or steam-heating system at a controlled back-pressure to obtain a thermal advantage compared with straightforward generation of electricity.

steam – saturated The properties of steam are such that, under a given atmospheric pressure, each steam pressure has a corresponding steam temperature when the steam is in contact with the water from which it is generated. This relationship only applies when the steam condition is dry (ie it contains no suspended moisture in the form of water droplets or mist). The steam is then said to be dry saturated.

steam – superheated When the steam is conducted away from contact with the water, it can be further heated by adding a degree of superheat in a superheater, usually to permit transport of the steam without causing

171

excessive condensation as the steam cools in the pipe system between the boiler and the point of use.

When superheated steam is used for heating, it has to be reduced in temperature to remove the superheat before the latent heat becomes available.

steam – total heat Contained in a given weight of steam is the sum of the latent heat of the dry fraction of the steam, the degree of superheat (if any) multiplied by the specific heat of same, and the sensible heat in the wet portion and of the water from which the steam is generated.

steam – wet Contains suspended moisture which will accumulate at low points in the pipelines and fittings. Such steam carries only a proportion of the latent heat capacity of a dry-steam system and imposes a greater load on the condensate collection system.

steam – wire-drawing Steam which expands through an orifice on a valve does no work and loses no heat energy; such expansion of steam is referred to as wire-drawing (or throttling). The process inevitably superheats the steam.

steam accumulator See *Accumulator – steam.*

steam air vents See *Air vents for steam.*

steam condensate meter See *Condensate meter.*

steam flashing See *Flash steam.*

steam generator – rapid See *Rapid steam generator.*

steam injection Refers to the direct injection of steam into a liquid for the purpose of heating that liquid. Process may be under thermostatic control (crude) or under manual control (usually wasteful).

Such injection is unsatisfactory for the following reasons: the steam dilutes the solution being heated; condensate is not recoverable; thermostatic or manual control methods tend to be crude, allowing an over-run of temperature and hence waste steam. When manual control is used, forgetfulness by the operator is not unknown when the steam is left on without need.

Application: temporary steam heating.
See also *Indirect steam heating.*

steam jetting Method of cleaning articles or plant by the use of steam at an appropriate pressure.

Generally, such cleaning should be discouraged, as it is likely to waste much steam. Other, more economical, cleaning methods should be considered.

steam leak Loss of steam through leakage at pipe fittings, steam traps, valves, process plant, etc depends on the operating pressure, the periods of steam use and the magnitude of the leak. The steam losses and the related costs can grow to alarming proportions. It is an inherent requirement of energy conservation to obviate or minimise leaks by daily monitoring of the steam system and prompt repair or replacement of the leaking components.

steam leak – cost of The table below shows the monthly loss of steam which arises from a *single point* of leakage of the given equivalent diameter at the stated steam pressure for a continuous process.

Equivalent diameter of the leak – nm (in)	Steam lost/month (kg)		Monthly cost of lost steam (£ sterling)	
	at 11 bar	at 7 bar	at 11 bar	at 7 bar
0.8 (1/32)	2,000	1,500	20	15
1.6 (1/16)	9,000	6,000	90	60
3.2 (1/8)	34,000	24,000	340	240
6.4 (¼)	140,000	96,000	1,400	960
9.6 (3/8)	300,000	215,000	3,000	2,150

Notes:
Process continuous: apply pro rata rates for other circumstances.
Generated cost of the steam leaks: taken at £10/1,000kg, given use of heavy fuel at 12.2p/litre (1981 typical price) and efficient boiler plant (operating at minimum 80 per cent efficiency).

steam leaks detector Handheld instrument for detection of steam leaks at distances of up to 15m from source. One such detector (Leakator by Dawe Instruments Ltd) detects the ultrasonic noise which high pressure steam leaks generate. The processed signal is used to vary the height of a vertical array of LEDs and the pitch of a tone fed to headphones. When maximum noise is indicated, a simple sighting system permits the operator to pinpoint the source of the leak. The directional properties of the instrument enable a leak to be located within 1m at a distance of 10m. Performance said to be independent of the noise of the leak and of the distance, within the operating range. *Energy Saving Potential:* can be very considerable, depending on extent of steam leakage(s). See *Steam leak – cost of.*
A good housekeeping tool for users of high pressure steam.

steam locking Occurs in trapped steam lines when the trap cannot open because the adjacent portion of the trapped system is full of live steam which must be condensed before trap can function. Any system in which steam can reach the steam trap in conjunction with condensate is liable to steam locking.
Such undesirable situation (which adversely affects heat transfer in the trapped vessel or pipeline) can be remedied by arranging a water seal ahead of the trap.

steam meter – integrating See *Integrating steam flow meter.*

steam meter – shuntflo See *Shuntflo steam meter.*

steam trap Designed to remove condensate and air. Available as thermodynamic, balanced-pressure, bimetallic, liquid expansion, impulse, ball float and bucket types.
Applications: where condensate has to be removed without losing steam, eg draining steam mains, tracer-lines, radiant panels, unit heaters, heating calorifiers and jacketed pipes.

steam trap efficiency monitor Checks whether a steam trap is wastefully blowing steam or not. A sensor chamber is close-coupled upstream of the trap with a cable running to a hand held indicator. One such monitor, the Spira-Tec (Spirax Sarc Ltd) allows rapid checking without the need to shut down. Defective traps can be positively identified, limiting maintenance to only those that require it.

steam trap – mechanical type Employs mechanical (as distinct from thermostatic) action to effect condensate clearance. Suitable for light or heavy condensate flow.
bucket model Actuated by an open or inverted bucket inside the valve body. Liable to airlocking, as air vents not normally incorporated. Very robustly constructed and unaffected by waterhammer; can be made of corrosion-resisting metal. Suitable with superheated steam and very high pressures.
Manual air venting encourages waste of steam. Under certain operating conditions, the inverted bucket trap can loose its water-seal; it will then blow steam (wastefully).
Discharges condensate at steam temperature as soon as it is formed. Working parts are simple. Traps can handle wide range of condensate loads from light to heavy.
Unsuitable for external location, where it may freeze up.
Application: process equipment, steam line trapping, etc.
Float model A float operates the trap discharge valve, opening it when full of condensate and closing it when trap body has emptied. Most types discharge entrapped air with the condensate by incorporating a thermostatic bellows-operated air vent in the upper part of the trap body.
Float can be damaged by waterhammer and corrosive condensate; float can leak and cause continuing malfunction.
Valve seat must be selected to suit operating pressure range, relative to size of float. Trap without thermostatic air release requires manual valve for release of air on start up to avoid air locking of the trap.
Application: on clean steam systems, such as space heating, where condensate load is quite large.

steam trap – thermostatic type Differentiates between steam and condensate by temperature difference, which actuates a temperature-sensitive (thermostatic) element carrying valve. Limited in principle due to steam and condensate at the valve inlet being at same temperature. Suitable for light condensate loads.

Balanced-pressure model Thermostatic element is a flexible bellows, which contracts and expands with temperature changes. Vulnerable to waterhammer and contaminated condensate. Unsuitable for use where superheat is present.
Application: process equipment of light steam duty, space-heating radiators, pipe main drain points, etc.

Liquid-expansion model Thermostatic element

is an enclosed liquid (oil) filled piston arrangement with packless gland operating on temperature difference. Can be used on superheated steam, and for discharge of condensate at low temperature. Discharge capacity greatest when the condensate is cool, as is case when starting a steam system. Not affected by waterhammer, vibration, steam pressure pulsation. Affected by corrosive condensate. Unsuitable for quickly changing condensate flow.
Applications: with superheated steam system, and where rate of condensate discharge is more or less constant as applies to certain process equipment.
metallic-expansion model Thermostatic element is a metal rod which moves the trap valve piston actuated by temperature changes. Smaller movement distance than liquid-filled trap for same temperature difference. Unaffected by waterhammer and corrosive condensate. Usually of relatively large dimensions.
Application: where there is risk of waterhammer and/or corrosive condensate or possibility of external damage.

steam trap – lifting trap Pistonless steam-operated pump in which a float operates a valve gear. When float at the bottom, water fills trap body by gravity flow through a non-return (check) valve. When trap has filled with water, float trips the valve gear and allows steam to enter into the top of the trap, driving the water out of the trap through a discharge pipe and a non-return valve into an elevated tank. Lifting trap can be used to remove condensate from a vessel under vacuum, but the trap must then exhaust into a vacuum.

steam whistle Audible signal conveyed by passing steam at pressure through a steam nozzle. Such practice is likely to be wasteful of steam and should be discarded wherever practicable in favour of an electrically actuated sound.

Stefan-Boltzman Law Relates to heat radiation and states that the heat radiated from a body is proportional to the fourth power of the difference between the absolute temperature of the radiating body and the absolute temperature of the surroundings, moderated by the emissivity of the radiating body.

step control Method by which the controlled variable is increased or reduced in definite increments (steps).
Applications: thermostatic control of banks of electric space heaters where different banks (sections) of heater elements are consecutively

energised or switched off to meet an increasing or decreasing heat demand. Control of direct expansion (DX) refrigeration equipment which comprises a number of refrigeration compressors and associated evaporators; the number of compressors in use is controlled in specific steps by switching in or out compressor units to meet a varying cooling demand.

step controller Electrical mechanism for effecting step control.
See also *Step control.*

sterilizer See *Autoclave.*

stoker – automatic See *Automatic stoker.*

stoker – chain grate See *Chain grate stoker.*

stoker-sprinkler See *Sprinkler stoker.*

stoker – underfeed See *Underfeed stoker.*

stop cock A 'shut-off' valve used for fluids circuit isolation and maintenance purposes.

storage heater See *Block storage heater.*

storage hopper See *Silo.*

stratification Formation of a temperature or density gradient in an enclosed liquid. For example, in a hot-water storage tank or cylinder, the lighter density heated water rises to the top and the heavier density cooler water falls towards the bottom. Thus temperature stratification is established across the height of liquid.
Particularly important in respect of storage of heated liquids, as stratification ensures that there are distinct layers of different densities, so that the hottest liquid will flow towards the upper section of the storage and can be drawn off as such, not withstanding that the lower parts of the liquid are cool and heating up (or cooling down).

strip curtain Heat retaining and draught reducing curtain usually manufactured from clear strips of PVC. Allows vehicles and pedestrians to pass through whilst remaining in position when not in use.
Applications: Loading bays where doors are left open; screening off sections of a factory; cold store doorways.

sublimation Change of state which takes place at very low temperatures directly from the solid to the gaseous state of a substance (bypassing the liquid state). Application in freeze drying.

submerged combustion Applies to gas-fired heating; in such systems the flame is submerged and burns below the liquid surface, the combustion products transferring directly to the contents of the tank in which the operation takes place.

Some of the heat is transferred to the walls of the gas-burner, but since this is immersed in the liquid, the heat is quickly transferred to this. The combustion products take the form of numerous very small bubbles which present a large total heat-transfer surface, so that the combustion products cool quickly to the temperature of the liquid in the tank; the combustion products and the water vapour produced leave the liquid surface together.
Applications: concentration of dilute solutions and crystallisation; tank heating, especially of corrosive liquids; water and effluent treatment. Figure 108 shows a submerged combustion burner of the nozzle-mixing type.

submersible pump See *Pump – submersible.*

sump – drainage Pit formed in an appropriate position in a plant room or service duct to collect intruding water from outside the building or duct. Requires a sump pump.

sump – oil Pit or depression in floor of boiler room or oil tank chamber into which surplus or leaks of oil are directed via suitable floor duct or channel. Serves additional purpose of housing a float control, which shuts off the oil supply to the boilers in the event of a major leakage.

sump pump Sturdy pump which is purpose-designed for location inside a damp environment in a sump and is provided with automatic float or electrode-actuated level controlled starter. Duty must be selected for the anticipated flow of water and the resistance (length and lift) of the discharge pipe.

sump – water Pit or depression in the floor of a service duct, plant room, etc to which water from pump glands, washing down, etc flows, induced by slope in floor or floor channels. Usually provided with sump or surface-located pump.

superconductivity Application of cyrogenic engineering principle in that many conductors of electricity diminish their resistance to the flow of electric current when maintained at very low temperature. Phenomenon then permits the passage of large currents through small size cables with little energy loss. Technique is currently emerging from the experimental and pilot stages; may eventually revolutionise design of large electrical machinery and of electricity transmission systems.

super heat See *Heat – super heat.*

superheated steam See *Steam – superheated.*

superheater Comprises of a set of inlet and outlet manifolds together with housing box and multilap elements which can be of welded or detachable design depending on application.

supervisory data centre Modular construction, microprocessor, programmable central control system. Twenty-four-hour and weekly programmes to control heating, ventilating and air conditioning of buildings and building complexes such as shopping centres, hospitals, etc. The centre can handle up to 1,000 address zones from single control point.
Applications: the system offers automatic optimisation of environmental control in buildings within prescribed limits by means of indoor/outdoor sensing of temperature, humidity, etc. Optional for building into the system are fire and security alarm initiation, automatic switching to standby equipment and fault correction systems.

support heat Required to raise the bed temperature from cold and to provide a trim heat facility during service. Various methods are used, including over-bed firing and an external form of direct fired air heater (oil- or gas-fired) for trim heating purposes.

support heat Required to raise the bed temperature from cold and to provide a trim heat facility during service. Various methods are used, including over-bed firing and an external form of direct fired air heater (oil- or gas-fired) for trim heating purposes.

surface heating Electric surface heating with resistance heating elements usually mounted in a glass cloth carrier for vessels, tanks, road tankers, etc.
Application: to protect contents of vessel from freezing and maintain temperatures of foods, liquids and chemicals.
Energy-saving potential: electric surface heating uses less power than steam or hot water jacketing – as little as a third. Heating is controlled by thermostats and only switched on when needed.

surface – selective See *Selective surface.*

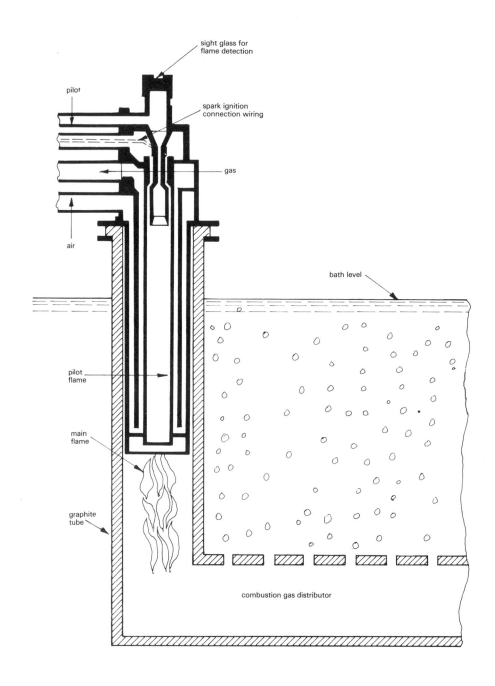

Figure 108 Application of submerged combustion by gas

surface water Water run-off from rooves of buildings, paved areas or ground surfaces (eg roads, car parks, hardstanding).

survey Investigation of an existing situation; eg ground conditions, buildings, utility services (sewers, telephones, electric cables, water pipes, etc).

suspended ceiling heating See *Ceiling heating – with false ceiling.*

suspended solids The solid particles in water. They can be filtered out, dried and weighed and are then expressed in terms of milligrams per litre.

swimming pool blanket A light-weight floating plastic blanket containing numerous sealed bubbles of air for use on outdoor and indoor swimming pools.
Applications: to increase water temperature and conserve energy on *outdoor* swimming pools. To conserve energy and improve pool hall environment for *indoor* swimming pools.
Energy-saving potential: pool heating costs can be reduced by up to 75 per cent, allowing existing plant to act only as a booster.

switch – flow See *Flow switch.*

switch-fuse Commonly applied to isolating switch for plant which embodies a suitable fuse carrier and fuse.
Object: to fuse-protect the particular item of plant.

synchronous motor See *Electric motors.*

tachometer Instrument for measuring the speed of rotation of shaft-driven machines.

tank farm Community of tanks within compound. Refers largely to oil storage tanks.

tar Viscous residue from distillation of materials, such as coal or timber. Black in colour. Coal tar yields a combustible tar oil; very viscous; requires expensive preheating and close combustion control; punishing on burner components.

tariff – block See *Block tariff.*

tariff – two-part See *Two-part tariff.*

tariff – white-meter See *White-meter tariff.*

tar sands See *Oil shale.*

telemetering Recording of variables at a central console fed with radio or telephone signals from remote detectors. *Typical Application:* central recording of widely spread pollution levels.

temperature A measure of subjective sensation (sense of appreciating hotness). Comparing two bodies which have been exposed to different intensity of heat, the hotter one is said to be at a higher temperature. Thus, the concept of temperature is essentially one of temperature *difference*. Scales have been devised to express this temperature difference (principally the Centigrade and the Fahrenheit temperature scales).

temperature – absolute See *Absolute temperature.*

temperature conversion Degrees Centigrade are converted into the equivalent degrees Fahrenheit by applying the multiplier 1.8 and adding 32.
Degrees Fahrenheit are converted into the equivalent degrees Centigrade by subtracting 32 and applying the divisor 1.8.

temperature – dry-bulb See *Dry-bulb temperature.*

temperature – effective See *Effective temperature.*

temperature – equivalent See *Equivalent temperature.*

temperature – globe See *Globe temperature.*

temperature indicating strip The strip adheres to the surface flat, and the indicators change from silver grey to black as the temperature rises. The change is virtually instantaneous and accuracy is 1 per cent of calibrated value. Such indicators can check almost any surface temperature and are water and oil resistant. Temperature indicated in °C or °F or both.

temperature indicator – digital See *Digital temperature indicator.*

temperature inversion Occurs when outside air is drawn from the top of the chimney around its periphery, whilst the hot gases rise up in the centre of the chimney; inversion causes smut deposition local to the chimney termination. Can occur at flue gas speeds below 5m/sec; much higher gas discharge velocities are necessary to avoid inversion with certainty.

temperature scales Measure the temperature using some suitable heat-related property, eg the expansion of mercury or alcohol in a thermometer or the tension in a spring.
Two temperature scales are in widespread use – Centigrade and Fahrenheit scales. Both are based on two fixed points, the lower one being that of melting ice and the upper one the steam generated by boiling water in contact with it under standard barometric pressure of 760mm of mercury (Hg).
The *Centigrade scale* has its fixed points at 0 and 100 respectively. The scale is divided into 100 equal parts; each is termed 1 degree Centigrade (°C).
The *Fahrenheit scale* has its fixed points at 32 and 212. The scale is divided into 180 equal parts; each is termed 1 degree Fahrenheit (°F). Each of the scales may be extended above the boiling point or below the freezing point. Temperatures below zero on either scale are denoted by a negative sign (eg -10°C means 10 Centigrade degrees below freezing; -10°F means $10 + 32 = 42$°F below freezing. (See figure 109.)

temperature – wet-bulb See *Wet-bulb temperature.*

temporary hardness Settles out of the water on heating and/or boiling. Causes scale deposition.

tetra silicates High-temperature heat-transfer fluids; tetra cresyl silicate commonly used with high-temperature liquid phase heating applications. See also *Heat transfer fluids.*

T

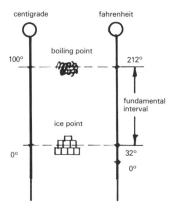

Figure 109 Centigrade & Fahrenheit thermometer scales

texture Relates to the distribution of individual grains within the material.

therm Unit of energy widely used in commerce and industry. Equals 100,000 Btu or its equivalent of 105.5MJ. Is likely to disappear with general acceptance of the SI system of units.

thermal comfort A sensation of human well-being within a particular environment. Depends mainly on the interaction of air temperature, relative humidity, radiated heat and air movement within the occupied space. Accepted that the prevailing air temperature is the most important thermal comfort factor in cool and cold climates, whilst humidity is likely to be of greater importance in very hot conditions.
The effective temperature scale and the equivalent temperature scale concept provide a measurement of physical conditions which define thermal sensations for the average person.
See *Effective temperature; Equivalent temperature.*

thermal conductivity (k) A measure of the ability of the material to transmit heat; expressed as W/m°C.

thermal fluid heater designed for temperatures of up to 350°C (662°F) for process applications using a high-temperature heat transfer fluid.
Application: where high process temperatures are required or where the low system pressures involved enable substantial savings in plant costs.

thermal image Heat pattern at surface level of buildings, insulated pipes, etc as recorded by infra-red camera. See also *infra-red camera; Thermography.*

thermal imagery Photography of heat patterns at surface level by infra-red camera and thermographic methods.

thermal insulation – Industrial Buildings Act 1957 In force in England and Scotland (not Northern Ireland). Stipulates that any new building or extension in any building subject to the Factory Acts must have its roof thermally insulated *if heating is specifically provided,* whenever a reasonable saving in fuel would ensue (unheated buildings, boiler houses and similar are excluded).
Rate of heat loss of roof (excluding lights, ventilators, etc) must not exceed 38 watts/m², based on – 1°C (30°F) outside temperature and the specified inside temperature measured at 1.52m above the working floor.

thermal insulation – loose-fill powder See *Loose-fill powder thermal insulation.*

thermal insulation – north light glazing Insulation unit which offers a second strong skin formed in PVC, typically about 2mm thick, which is fitted to the exterior of roof-lights to form a double-glazed unit; application by special adhesive. One such unit (Northlite I.V. 40/60 – Northlite Insulation Services Ltd) is said to improve U valve of the glazing from 7.54 w/m²/°C to 2.53 w/m²/°C. For factory premises in use 12 hours/5 days a week, 36 week heating season and heated by oil, a payback period of about two years is quoted.

thermal insulation – rigid foam See *Rigid foam thermal insulation.*

thermal pipe insulation – protection Cement bandage applied over the thermal insulation material. One such product is Rok-Rap (Evode Ltd). To use, soak in water and wrap over insulation to obtain a rock-hard mechanical protection. Suitable for all types of insulation material, including asbestos; claimed to resist corrosion, fire and chemical attack.

thermal resistance (r) The product of thermal resistivity (1/k) and thickness of the material (m); expressed as m²° C/W.

thermal resistivity (1/k) The reciprocal of thermal conductivity; expressed as m°C/W.

thermal storage electric system Operates with lower priced off-peak electricity by heating water using electrode boilers during the off-peak period and storing the water for use during 'working period(s)'. Required storage vessels are generally relatively bulky – must be thermally insulated to highest practicable efficiency.

In use, the storage vessels connect to a pumped circulation which distributes the heat conventionally.

thermal transmittance (u) A property of an element of a building fabric (or structure) which consists of given thicknesses of material; it is a measure of its ability to transmit heat under conditions of steady flow. Defined as the quantity of heat that will flow through unit area, in unit time, per unit difference in temperature between the inside and outside environment. Calculated as the reciprocal of the sum of the resistances of each layer of the construction and the resistances of the inner and outer surfaces and of any air space or cavity; expressed as $W/m^2°C$.

thermal wall See *Trombe wall.*

thermocouples Consist of two wires of dissimilar metal joined together at *both* ends to form two junctions; used to detect, measure and relay temperature readings, particularly from remote locations. The junction at the point of the hot-temperature measurement is termed the *hot junction;* that at the lower temperature is the *cold junction.* They must be so located that a temperature difference exists between them at the operative time which causes the generation of an electro-motive force (emf) in the loop which actuates an appropriately calibrated associated measurement apparatus.

thermodynamics – first law See *First Law of Thermodynamics.*

thermodynamics – second law See *Second Law of Thermodynamics.*

thermography Infra-red imagery using purpose-made camera with nitrogen-cooled silicon-coated lens. Equipment detects infra-red radiation from buildings, pipelines and other objects and displays a continuous picture on a monitor or television screen. In these thermal pictures, the coldest areas show black, gradually lightening until the hottest areas show white.

Applications: energy-conservation surveys, routine maintenance of insulated buildings, pipelines, etc and of electrical installation and for detection of leaks and guidance to emergency repairs.

Thermography may be by ground-level or by air-borne survey.

thermography – air-borne survey Carried out from specially adapted aircraft with downwards directed infra-red camera, flying at 500m to 750m above the ground.

Useful for survey of large sites (or groups of sites) and of district-heating distribution networks to establish major energy losses. Of considerable importance in large-scale energy-conservation schemes to pinpoint major areas of concern, which will then be subject to local inspections. Can be employed to check on the standard of thermal insulation (of, say, roofs) relative to specification without disturbing the installed construction.

1981 cost of air-borne thermographic survey in the order of £20,000 ($40,000) per week.

thermography – ground-level survey Uses hand-held or tripod-mounted infra-red camera. Has wide application in energy conservation and audits. Provides accurate close-up thermal images. Can be employed to survey buildings and installations without disturbing production. Particularly useful for inspecting the condition of the refractory and insulation linings of chimneys and production plants whilst plant remains in use; can inspect the inside of electrical panels without opening up and will indicate area(s) of overheating which points to a malfunction and requires remedial action.

Can establish U-values of building elements (given certain associated information). Used widely in Europe for monitoring and emergency scanning of district-heating underground mains and of other buried pipes and cables. 1981 cost of thermographic ground-level survey and report in order of £600 ($1,200) per day.

thermography – survey May be taken at ground level or be air-borne, using infra-red camera.

thermo-mechanical The process of thermal expansion or contraction of a temperature-sensitive metal or fluid employed to actuate the mechanism of a control device on a change of temperature.

thermostat A sensor, electrical or non-electrical, measuring temperature fluctuations and activating a related control function.

thermostat – differential Automatic device which senses the difference in temperatures among several related components in a system, such as a solar collector, its store and the heat application, and directs the heat flow as required or specified.

thermostat – room See *Room thermostat.*

thermostatic radiator valves Self-acting liquid-filled radiator valves sensitive to ambient temperature and in sizes to fit most pipe diameters for gravity-circulated flow, pumped systems and steam systems. Models available with integral and remote sensors.
Application: modulating control of individual radiators acording to preset ambient temperature requirement.

thermosyphon The hydraulic system in which a fluid circulation is caused by temperature (thus density) differences in the fluid.
(Others of only immediate significance are explained in the text as they occur.)

three-phase circuit Applies to an electrical installation where the alternating current is divided into three branches, or phases. Generally used with non-domestic and the larger commercial/industrial/lift installations.

throwaway air filter See *Air filter – throwaway type.*

thyristors drive Controls the speed of d.c. motor by armature voltage variation. Using solid-state adjustable speed control, this type of drive can meet the performance requirements for most industrial applications. The use of d.c. motors is inhibited by the high capital and maintenance cost of these; the main advantage is that the overall cost of the d.c. drive is less than the corresponding control for an equivalent a.c. motor.

timber kiln See *Kiln.*

timelag Interval of time between action and noticeable effect, eg there will be a timelag between firing the boilers in a heating system and the temperature rise within the building; conversely, there will be a timelag between switching off the boilers and the building cooling down.

timers May be manually operated clockwork or motor driven timers; typically the scale of times range from 3 min to 2 h with electrical switching contacts rated at 12 amps at 250V. On the motor-driven timer, the range of time

scales is larger, and 16-amp rating is also available.

ton of refrigeration Term used to measure the cooling capacity of refrigeration plant. One ton is equivalent to a capacity of 3.15 kw (12,000 Btu/hr) (S.I. Units)

total energy system Supplies all the energy requirements of a complex or premises by the in-house generation of electricity and recovery of waste heat from the turbines, diesel generators, etc, by back-pressure steam techniques, jacket cooling, etc. May include supplementary heat generator(s).
See also *Combined heat and power.*

total heat See *Heat – total.*

town gas Manufactured by the carbonisation of coal in gas retorts. Has calorific value of $20MJ/m^3$. Supplied to generality of gas consumers. Gas is toxic when inhaled.
In the UK, town gas has been largely superceded by natural gas won from the North Sea.
See also *Natural gas.*

transformer Equipment used with electricity transmission systems. Has two separate windings and permits the downgrading of voltage from the economic high transmission voltage to the lower voltage required for the consumer equipment.
Transformation involves a heat loss and hence the transformer must be provided with means of cooling to dissipate this energy loss.

transistor Semi-conducting device which incorporates three or more electrodes. Used with electronic (control) equipment.

transmissibility The amount of vibratory force that is transferred to the structure through an isolator, expressed as a percentage of the total force applied.

transparent reflective surface A surface coating which allows short-wave radiation to pass through it while infra-red (thermal) radiation is reflected.

trap See *Steam trap.*

trap – condensate See *Trap – sanitary.*

trap – sanitary U-bend which provides a water-seal between the sanitary appliance(s) and the drain or sewer. Traps must be provided also at the point where condensate drains from air-conditioning and similar equipment enter the main disposal system (whether rain-water or soil).

trapped gully See *Gully – trapped.*

trombe wall(s) Relates to a passive solar-energy-collection method which employs thermally massive wall construction for the storage of solar energy. The trombe wall is placed between the living space of a house and the external glazed suntrap. The wall collects solar radiation and stores the heat energy within its mass; this is then reradiated through the back of the wall and/or by convection movement of the cooler room air along the warm face of the trombe wall by the natural buoyancy of the air. This 'thermosyphonic' effect is created on the face of the wall exposed to the sun by locating a glazed screen about 50mm distant from the sun-facing side of the wall, encouraging the desired air circulation by forming openings near the top and close to the bottom of the trombe wall. The air circulates through the openings and the space between trombe wall and sun screen, absorbing heat and being raised in temperature. (See figure 110.)

The trombe-wall method of heating can be controlled in several ways: the thermal mass of the wall is selected to provide sufficient timelag before reradiation occurs; manually (or if the expense is considered justified, automatically) operated dampers regulate the air flow through the openings; sun-blinds are manipulated during the summer to prevent overheating.

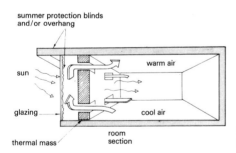

Figure 110 Trombe or thermal wall application

tungsten lamp Incorporates a metallic element used as filament. In use, inefficient relative to fluorescent lighting. Now used mainly for domestic and special effects lighting.

turbine Directs the flow of a liquid or gas at high pressure to provide rotational movement. Turbines are equipped with blades, vanes,

nozzles or similar to effect efficient transformation of energy.

Steam turbines are employed for electricity generation in major power stations. Gas turbines are used for smaller size peak-load or localised electricity generation.

turbine – back pressure Outlet pressure is raised, so that the turbine operates against a back-pressure at the outlet. Back-pressure steam piped to process or general heating application. Small combined heat and power back-pressure systems yield between 6 to 8 units of heat for each unit of electricity sacrificed. (See figures 112 and 113.)
Application: factories of adequate size to justify independent electricity generation and with a balanced heat requirement.

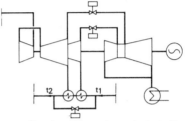

a. Non-reheat DH-extraction condensing turbine

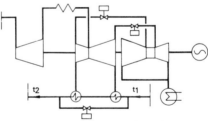

b. Reheat DH-extraction condensing turbine

Figure 111 Non-reheat types of extraction – condensing turbines

turbine – condensing Operated with the object of achieving the highest possible electric power generation efficiency by employing a vacuum on the outlet side; practicable to achieve a pressure of 4kPa on the condensing side of turbine. (See figure 111.)
Application: main power stations.

turbulators Metal strips formed into opposing 30°, 45° and 60° bands. When inserted into boiler tubes, they create turbulence in the hot gases, increasing the rate of transfer of heat to the boiler water. By careful grading of length,

turbulators

lower tubes are made more effective improving the distribution of heat.

Applications: designed to suit fire tube hot-water and steam boilers. Can also be used in tube heat exchangers, after burners, waste boilers and warm air heaters.

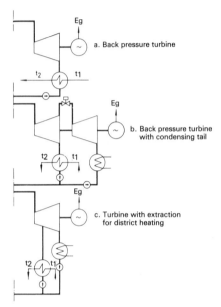

Figure 112 Back pressure turbine arrangement for district heating

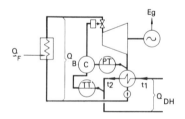

Figure 113 District heating back pressure turbine

turbulent flow A confused state of airflow that may cause noise to be generated inside, for example, a ductwork system.

turn-down ratio Relates to boiler and furnace firing equipment. Specifies the ratio to which this can be adjusted when firing the fuel, eg 100 per cent provides full input; 33 per cent, one-third input. Wide range of turn-down implies flexibility and tends to improve boiler efficiency when serving a fluctuating heat or process load.

turning vanes Metal blades incorporated into air ducts at changes of direction to promote smooth and even air flow through such fittings into the adjoining sections of duct.

two-duct air-conditioning system Utilises two separate air supply duct systems, one carrying warm air and the other cooled air into a mixing box, from which the correctly blended air is carried to the conditioned space.
System now generally out of favour due to high cost and space requirements.

two-part tariff Comprises a fixed-charge element (levied per week/month/quarter/ **annum) or proportional to the maximum elec** · trical demand or to the total connected load, plus a charge per unit consumed.

two-position control See *Automatic control – two position.*

ultrasonic Inaudible (to the human ear) pulsations of high-frequency sound; typically 28,000 cycles/sec.

ultrasonic air and steam leak detector
Designed to detect faulty steam traps, steam leaks, air and gas leaks from both pressure and vacuum systems. Audio and visual indicators. Battery operated for portability.
Application: to detect faulty steam traps and minute steam air or gas leaks from pressure and vacuum systems.

ultrasonic boiler cleaning The ultrasonic pulses are transmitted to the boiler water through a length of pipe. The metal disc transmitter operates at 28,000 cycles/sec frequency and pulsates about 3 times/sec for a duration of 1/1,000 sec.
Application: satisfactory results have been reportedly achieved with shell boilers steaming on a supply of chemically untreated water of average hardness at pressures up to 30 bar. Normal operation permits the scale to build up to eggshell thickness; it is then broken off in patches which drop to a scale-collection pocket.
Caution is essential, in that the ultrasonic power must be limited to prevent the destruction of the protective mill scale on the boiler steel plates.

ultrasonic measurement of oil level in switch-gear Portable, battery-powered measuring instrument (Londalevel by Londex Limited, London) transmits an ultrasonic beam through tank of oil-filled switchgear and detects internal oil level within 10mm. Operates wholly outside vessel being checked, and the switchgear can remain operational whilst the oil levels are monitored. There have been a number of damaging explosions caused by the lack (loss of oil in switchgear equipment); the availability of the ultrasonic equipment should lead to increased safety.

ultra-violet A band of electromagnetic wavelengths adjacent to the visible violet (0.10 nm + 0.38 nm) of particular interest in solar-energy application, since various unprotected materials change in colour due to reception of ultra-violet radiation; if the materials are unstable, they are then likely to suffer failure due to decay or fracture.

underfeed stoker Screws the fuel into a retort, where it is then burnt from the top of the bed. Usually arranged to operate at a selected choice of available fuel feed speeds. (See figure 114 for such stoker adapted to firewood waste.)

under-floor heating See *Floor heating.*

uniformity coefficient A term often used to define the grading of filter sands. It is the ratio between the aperture openings passing 60 per cent and 10 per cent by weight of the sand sample.

units – S.I. See *S.I. (units).*

unloading equipment Relates to air compressors and is associated with the automatic air governor of same. When the air in the air receiver of the compressor has reached a predetermined pressure, the automatic air governor passes compressed air into the unloading cylinders, which are fitted above the compressor suction valve. When the pressure falls to the predetermined lower limit of the air

U

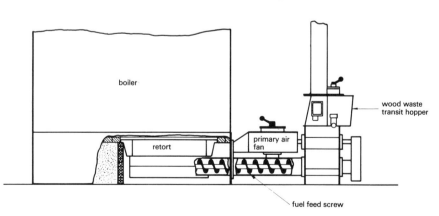

Figure 114 Underfeed worm stoker

governor adjustment, the air supply to the unloading cylinders is cut off by the air governor.

unplasticised PVC pressure pipes, joints and pipe fittings Injection moulded pipe fittings made from unplasticised polyvinyl chloride pressure pipes.

urea formaldehyde foam Widely used for insulating existing brick or timber cavity walls. Can be conveniently injected into such cavities through small holes on the inside or outside of the wall and sets without any appreciable heat change.
The material is unaffected by dilute acids, alkalis, oils or solvents. It neither rots nor shows any sign of aging. It is incombustible and simply shrivels up when exposed to high temperatures. The material is greasy in nature, so that, while it presents no barrier to water vapour, it tends to repel liquid water by capillary action. Use of material is suspected of posing a health hazard to building occupants.
USER (utilisation) factor (solar collection) A factor employed in calculating the actual benefit derived from a solar hot-water supply system based upon consumer habits.

u-value calculation The coefficient U (thermal transmittance) of a composite building structure or element is expressed as:

$$U = 1/(R_{si} + R_{so} + R_{as} + R_1 + R_2 + R_3)$$

R_{si} = internal surface resistance
R_{so} = external surface resistance
R_{as} = resistance of air space or cavity
    forming part of the building element
R_1 = resistance of first layer of material
R_2 = resistance of second layer of material
R_3 = resistance of third layer of material
    (if present)

Tabulated information on the above parameters and coefficients is available from various publications, including those of the Chartered Institution of Building Services, Building Research Establishment (UK), as well as from numerous technical books.

Example:
Calculate the U-value of a wall comprising:
outer face – unrendered.
brickwork – clay @ 1700 kg/m³ – 100mm thick.
cavity (sealed – air space) – 50mm wide.
internal face formed in lightweight block – 500kg/m³ – 100mm thick.
plaster finish on internal wall – 10mm.

R_{si} = 0,123m²°C/W

R_{so} = 0.550
R_{12} = 0.180
External wall leaf (R_1) = 0.10/0.84 = 0.119
Internal wall leaf (R_2) = 0.10/0.19 = 0.526
Plaster finish (R_3) = 0.01/0.19   = 0.053
                          1.056

U-value = 1/1.056 = 0.95 W/m²°C

Note: coefficients taken from BRE publications.

U-value definition Equates with thermal transmittance. A measure of the ability of the different building elements (walls, floors, roofs, windows) to conduct heat out of the building; the greater the U-value, the greater the heat loss through the building element. The total heat loss through the building fabric is the summation of all the elemental U-values and the related externally exposed areas, multiplied by the difference in temperature between the internal and external environment. The U-value of a structure (or building element) varies to some extent about a mean value from one situation to another; it is affected by moisture content, wind speed, internal conditions, etc. The standard U-values are calculated from the resistances of the building components (based, in turn, on standard assumptions concerning moisture contents of materials, rates of heat transfer to surfaces by convection and radiation, and airflow rates in ventilated air spaces). Some allowance is also made for certain heat bridging through the structure. The standard assumptions are essentially based on practical conditions, although these cannot apply in all circumstances of building locations. Measured U-values are not accepted as standard, because the conditions of the measurement only seldom agree precisely with the standard assumptions.

valve – balancing See *Balancing valve system.*

valve – ball See *Ball valve.*

valve – bleed Arranged to allow a relatively small quantity of fluid to flow from the system.
Application: cooling pond bleed to limit accumulation of dissolved solids in the pond and similar.

valve – blowdown Specially adapted for control of quantity of high-temperature water being discharged periodically (blown down) from steam boiler to limit accumulation of dissolved solids within steam boiler. Strongly constructed to withstand high pressures. May be operated manually by extended handle or automatically by more or less sophisticated blowdown control system. Designed to safeguard boiler operator.
Application: steam boiler plant.

valve – butterfly See *Butterfly valve.*

valve – constant-flow See *Constant-flow valves.*

valve – diaphragm Controls flow of liquid through valve body by a diaphragm, so that there is no gland through which liquid might leak. Diaphragm is replaceable on failure. Different diaphragm materials available for a particular valve; selected to suit specific liquid being handled by valve. Most important that correct diaphragm is specified.
Applications: oil, chemical and general process industry.

valve – double-regulating Provides means of a fixed permanent adjustment of the valve plus ability of closing the valve completely to provide shut-off.

valve – gate Controls flow of fluid through valve body by vertical movement of a gate attached to (screwed on to) the valve spindle.
Applications: liquid flow isolation under conditions of relatively low head and temperature; eg cold water down services, low-pressure hot-water heating systems, etc.

valve – gland Has wheel head protruding above the gland.
Application: drain valve and similar use.

valve – globe Controls flow of fluid (liquid or gas/vapour) by controlling action of a jumper (usually fitted with a washer) which acts on a valve seat. Fluid enters below the seat and leaves above, after passing through the seating.
Application: high-pressure mains water systems (eg stopcocks), steam systems, etc.

valve – lockshield Has, instead of conventional operating wheel, a metal shield over the valve head. Shield usually screwed on to valve body to inhibit interference with the established valve setting.
Application: regulating valve in heating and hot-water supply systems.

V

valve – non-return See *Non-return valve.*

valve – parallel slide Controls the flow of fluid through valve body by vertical movement of a slide which moves within parallel guides.
Application: liquid flow and steam isolation under more arduous conditions, such as apply in high-pressure hot water and in certain process systems.

valve – pit Housing for valve(s) associated with underground piped services.
Valve pits for underground district-heating installations require most stringent care to ensure water-tightness. Some preinsulated pipe

systems embody preinsulated valves and use simplified valve actuating chambers to eliminate the potential failure risk associated with conventional valve pits. (See figure 83.)

valve – pressure-reducing See *Pressure-reducing valve.*

valve – reversing See *reversing valve.*

valve – rotary See *Rotary valve.*

valve – solenoid See *Solenoid valve.*

valve – squeeze Handles abrasive and corrosive fluids, including solids in suspension at pressures typically up to 12 bar. Operates with vulcanised rubber sleeve, wich is reinforced with woven fabric. Sleeve returns to original shape when the valve reopens. Larger valves are strengthened with steel cord reinforcement. Complete isolation is achieved when the valve is pinched. Typically available in sizes of between 40mm and 500mm.
May be provided with built-in rubber flanges to eliminate need for gaskets between pipe flange and valve. Available closing mechanisms include manual (handwheel), hydraulic, pneumatic and motorised systems.

valve – thermostatic radiator See *Thermostatic radiator valve.*

valve – Y-type Has its valve spindle inclined at an angle into the direction of flow. Offers the minimum resistance to liquid flow consistent with good stop-valve design.
Available as simple or double-regulating valve pattern; also with drain/vent cocks or with blank plugs.
Available with a backseating machined feature on the valve stem which enables the valve to be packed under pressure when fully open.

vane anemometer Portable air-measurement instrument; essentially a windmill which is free to rotate inside a metal guard and a centrally located dial which is graduated in metres per second, over which moves a pointer gear driven by the windmill. May incorporate stop-and-start and set-to-zero mechanism. Small and compact.
Application: for measurement of air flow at air registers and other locations where the air velocity is relatively low.

vapo(u)r Gas which may be condensed by increase of pressure *without* reduction of temperature (unlike a permanent gas); ie it can be fairly easily liquified. Example: steam (water vapo(u)r.

vapo(u)rising burner Oil burner used in small output domestic heat applications. Operates with vapo(u)rising oil (of low viscosity) vapo(u)rised inside a pot in which the vapo(u)r is mixed with the correct quantity of air to achieve good combustion. May operate with natural or fan-assisted draught. Such burners tend to be maintenance intensive and subject to adverse draughts. No longer widely popular.

vapo(u)r pressure Pressure exerted by a vapo(u)r on its own or in a mixture of vapo(u)rs or gases. According to Dalton's Law, the pressure which a vapo(u)r exerts is almost independent of the presence of any other gas or vapo(u)r in the mixture.

vapo(u)r pressure – saturated A vapo(u)r which is in contact with the liquid from which it was formed (eg steam inside a boiler) is said to be saturated and contains the maximum liquid it can absorb at that pressure. Reduction in pressure leads to condensation. The pressure at that condition is the saturated vapo(u)r pressure.

vapo(u)r pressure – unsaturated A vapo(u)r out of contact with its liquid base is likely to be unsaturated; it can absorb further moisture (eg vapo(u)r in the atmosphere). The pressure at that condition is the unsaturated vapo(u)r pressure.

vapo(u)r seal Continuous seal which provides an efficient barrier to the transmission of moisture. Most effective materials for vapo(u)r seals are polythene sheeting (heavy gauge), metal foil and metal sheet. Also used (though increasingly out of favour) are several layers of bitumenous emulsion of hot bitumen. The vapo(u)r seal is placed across the *warm* side of the insulation (the inner face will be the cold side). If at all practicable, the inner wall coating should embody some porosity to permit small amounts of moisture (which might pass through minor defects in the vapo(u)r seal) to migrate, rather than be entrapped and cause insulation failure.

variable speed drive Variable frequency inverter for driving standard induction motors at variable speed. Such operation of fan, pump and compressor systems improves efficiency and control instead of damping or throttling; eg a 100kw pump running at 80 per cent throughput will absorb approximately 85kw if throttled back or approximately 50kw if run at variable speed. Two years is a common pay-back period; may be less for the larger systems.

variable speed drive for a.c. motors Can be used with existing a.c. motors to upgrade an existing installation.
Application: variable speed control of pumps and fans from 5kw to 20kw.
Energy-saving potential: depending on application, motor running costs can be reduced by approximately 20 per cent, giving an average payback period of two to three years.

variable speed drives Adjust the speed of elec-tric motors to follow the imposed load. Use can result in considerable energy savings. Largest potential in drives up to 200kw; within this range, the most common variable speed drives are: eddy current, d.c. thyristor controlled and a.c. invertor types.

variable-volume air-conditioning system All-air system which operates on the principle of automatically varying the quantity of conditioned air supplied to the space to match the cooling requirement. Can utilise the outside air for 'free cooling' when the air temperature is at, or below, 13°C (55°F). Incorporates arrangement of master being controlled by a thermostat in the controlled space. Main supply fan automatically compensates for changes in air volume requirement.
Most systems are located within suspended ceilings. Minimises energy consumption. Capital cost relatively high.
Application: buildings with deep internal spaces and low perimeter losses. It is usually desirable to have some background heating at the perimeter.

V.A.V. system See *Variable-volume air-conditioning system.*

vee-type air filter See *Air filter – vee-type.*

velometer General purpose air-velocity measuring set for portable applications offering ten ranges from 0-300 ft/min to 0-10,000 ft/min + 0 to 1 and 0 to 10 in of water gauge static pressure. A metric equiv-alent is available.
Application: measurement or detection of moving air in heating and ventilating systems.

ventilation Relates to supply and extraction of air from an environment. May include provision for humidification, heating,

infiltration and air distribution.

ventilated ceiling Purpose-made (most commonly) metal ceiling suspended below the soffit of the structural ceiling of the space adapted for the downward displacement of ventilation air. Air outlets may be perforations or adjustable slots located in module config-uration. Ceiling incorporates acoustic and thermal insulation. For best effect, integrated with flush moulded light fittings.
Applications: clean rooms, computer suites, etc.

Venturi meter Practical application of Bernoulli's Theorem to the measurement of fluid flow. In basic form, the meter consists of a short length of pipe which tapers to a narrow throat in the middle. Detector tubes enter at the throat and at the ends and measure the pressure of the fluid at these sections. As the liquid flows through the throat, the velocity must increase due to the reduced area; consequently, the pressure will be reduced. By applying Bernoulli's Theorem to the readings taken at the enlarged end and to the throat, the quantity of water flow can be measured and calculated. (See figure 115.)
Application: more or less sophisticated meters for the measurement of liquid and gas flows are based on the Venturi meter principle. There will inevitably be a loss of head in the meter between the enlarged end and the throat; this can lead to significant errors at low velocities; hence this meter is not suitable for the accurate measurement of fluid flow at low velocity.

vermiculite The geological name given to a group of hydrated laminar minerals, which are aluminium magnesium silicates and resemble mica in appearance. The sintering temperature of vermiculite is about 1,260°C (2,300°F) and its melting point is 1,315°C (2,400°F). The material is found in many parts of the world, one of the main supply regions being the Palabora deposits in the north-east Transvaal, where it is mined in open cast methods. Rock and other impurities are removed, and the crude ore is crushed and graded.
Vermiculite, in various guises, is used for thermal insulation; eg in the form of granules (loft insulation), screed (roof insulation)

Versatemp air-conditioning system Proprietary system which functions without a central chiller. Utilises individual air-conditioning cabinets which are located within the space and are room thermostat controlled.
Each unit incorporates a refrigeration

compressor, which operates in conjunction with a two-pipe circulation connected to a centrally located heat exchanger and indirect cooling tower via suitable control valving. The water circulation is close to room temperature and hence does not require thermal insulation. Each unit is self-actuating between preset temperatures for heating or cooling. Heat-recovery feature permits heat from zone being cooled to be used for zone which requires heating (eg autumn situation where south face requires cooling and north face heating). Low-cost system when considering overall costs. Noticeable noise when changing from heating mode to cooling mode.
Application: commercial premises.

vibration isolation Any of several means of reducing the transfer of vibrational force from the mounted equipment to the supporting structure, or vice versa.

viscosity The resistance to flow between adjacent layers of a fluid. A fluid which flows steadily in a conduit with smooth sides will move at a velocity which varies over the cross-section of the conduit. The layers of fluid adjoining the perimeter will be at rest; layers adjoining these will move at a low velocity; there will be a gradual increase in the velocity of the layers of the fluid, reaching a maximum when the centre of the conduit is approached. Thus, since two adjacent layers of the fluid move at different velocities, there will be a resistance to flow between them – *the viscosity*. Viscous resistance of a fluid is analogous to the shear resistance of a sólid, probably due to molecular attraction acting on planes inclined to the direction of flow. The frictional resistance of a fluid in a (rough) pipe is caused

by its viscosity, as the rough sides of the pipe cause cross-currents and eddies, the energy of which is ultimately destroyed by viscosity. Actual viscosity of a particular fluid depends on its coefficient of viscosity, on the temp-erature, its density, its velocity and on the size of the conduit. A viscous resistance is inde-pendent of the material of which the pipes are constructed. Dynamic viscosity is measured in stokes (cm^2/sec); kinematic viscosity in poise (dyne/sec/cm^2).

viscosity – coefficient of The relationship between the viscous stress and the angle of distortion (due to viscosity).
Tabulated information widely available concerning the coefficients of viscosity of fluids.-

viscosity – kinematic The relationship between the coefficient of viscosity and the density of the fluid.

vitrification Term used with refractories; relates to the conversion of a material under heat into a glass or glass-like substance with increased hardness and brittleness. This partic-ular characteristic of a refractory brick is of great importance in its development of high strength and hardness.

voltage stabilizer See *Automatic voltage stabiliser.*

volumetric efficiency Relates to reciprocating compressors, and expresses the loss of working stroke due to the high-pressure gas which is left over in the clearance spaces of the compressor at the end of the discharge stroke and which must reexpand to the suction inlet pressure

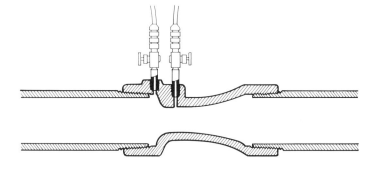

Figure 115 Barco-Aeroquip Venturi system for measuring water through velocities

before a fresh charge of gas can enter the compressor. Clearance spaces tend to be about 4 per cent to 6 per cent of total swept volume. Loss of useful working stroke varies with the compression ratio. Assuming R22 gas and total clearance volume of 7 per cent, the typical *volumetric efficiency* will fall from 1(100 per cent) at a compression ratio of 1 to 0.5 (50 per cent) at a ratio of 12.

wall bush Special rubberised fitting employed to waterproof the interface between building pipe entry and the entering service pipes.

wall-hung boiler Compact residential heating boiler arranged for being supported in a wall-hung position. Maximum rating about 20 kW. Small water content. Available for conventional or balanced flue. Neat appearance and saves floor space. Package may include circulating pump, controls and programmer.

warm air furnace Appliance for direct heating of warm air. Comprises within one assembly, burner, combustion system, ventilation fan, air filter, controls and means of air distribution. May discharge air directly from integral outlet louvres or nozzles or may be connected to distribution ducts off integral flanged connections.
Appliance is usually connected to a suitable chimney to evacuate the combustion products. Free standing on floor, suspended above the floor or enclosed in a separate plant area. Fuelled by gas, oil or solid fuel; some furnaces are adapted to the burning of waste or sump oil.
Operational controls are usually two thermostats and a time clock. The room thermostat switches the burner off when the heat demand has been satisfied; a further internal thermostat then switches the air fan off when the temperature within the assembly has dropped to a specified setting. This thermostat will delay start of air fan until the air has warmed up. It is possible to modify the controls to permit the air fan to operate for the circulation of air (such as in summer) when the heating requirement has been satisfied.
Applications: best suited to the heating of large open-plan areas, such as factories, warehouses, aircraft hangers, etc. Enjoys some popularity in residential heating, particularly in North America.

waste bulk density See *Bulk density – waste.*

waste compaction Reduces the volume of refuse by as much as 8 to 1, though most commercially available compactors compact to about 4 to 1.
Application: for dry and wet material; where a high degree of volume reduction is required; used in shopping centres, canteens, etc.

waste-derived fuel (RDF) One end product of a waste-recycling process. Sufficiently dense to permit transportation and handling. Calorific value depends on constituents. Figure 90 is the flowsheet for the manufacture of refuse-derived fuel.

waste-heat boiler Designed especially to operate with the waste gases discharged from incinerators and similar equipment, the boiler being inserted between the incinerator and the chimney.
Waste-heat boiler may generate steam or hot water, as required.
The smoke tubes of a waste-heat boiler are usually of larger diameter than those fitted to conventional boilers, being related to the contaminated gas that may be handled by the boiler. There must be good access for cleaning of the smoke tubes and the other parts of the boiler and combustion system.
When the heat demand is seasonal, the gases are either by-passed to the chimney out of season or the heat is dissipated by blowing off steam (undesirable because of loss of treated water and possible nuisance) or by passage through a heat dissipator. The use of a gas by-pass implies the provision of suitable (expensive) chimney internal insulation lining.

waste oil burner Usually can burn all grades of oil, including waste oil from manufacturing and vehicle sumps, burner operating on principle of emulsification of oil/air mix.

waste pulverising machine Swinging beater size reduction hammer mill capable of reducing process scrap; eg wood waste from furniture manufacture to boiler fuel.
Application: producing boiler fuel from waste.

waste recycling – direct Return of manufacturing waste to the original process; eg broken or misshapen glassware returned to the furnace for remelting and reuse.

waste recycling – indirect Some properties of the waste are recycled: eg use of wood waste in chipboard manufacture; use made of the heat generated during incineration of wood waste, tyres, etc; use of pulverised fuel ash from power stations for civil-engineering purposes; conversion of waste plastic containers into coarser products, such as fence posts.

water conditioner – electric See *Electric water conditioner.*

water conditioner – magnetic See *Magnetic water conditioning.*

water equivalent (heat capacity) The mass of water which requires the same quantity of heat to raise the temperature through 1 degree as the body requires; eg the water equivalent of 9g of

water flow restrictor

iron (specific heat 0.119) is about 1g (specific heat of water taken as 1.0).
If M = mass of the body
  s = specific heat of the material (body),
then water equivalent (or heat capacity) = Ms. See also *Heat capacity*.

water flow restrictor A device fitted within the body of a water draw-off tap which reduces flow to the required level. For use in hot and cold water installations in schools, public buildings, institutions, factory washrooms, offices, etc. Will reduce water flow by 50-75 per cent.

water hammer Vibration (usually accompanied by loud noise) in a water or steam system. May be caused by a variety of disturbances, such as presence of air pockets, chattering valve parts, loose ball valves, etc.

water hammer – elimination of Water hammer is easily diagnosed, because of the accompaying commotion (vibration and noise). In steam systems, water hammer tends to be at its fiercest when the steam is first turned on and the pipes are not clear of condensate. The point of maximum water hammer can, in most cases, be accurately located (if necessary) by observation over a prolonged period of time. Water hammer can be eliminated by ensuring that all steam and condensate pipes are correctly graded to permit flow of water with the steam flow; all air is eliminated; vent points and steam traps are maintained in good order; drain pockets are fitted to the low points of the steam pipes ahead of the traps; all reducing fittings in the pipe system are of the eccentric type to permit smooth flow.

water hammer – in steam systems Caused usually by contact between live steam and water (eg condensate), setting up a high-velocity wave motion. Shock waves are generated at obstructions (eg changes in direction of the pipe) when the slugs of water collide with the metal and are brought to rest with considerable shock. The consequent destructive effect may be great, resulting in cracked fittings, fracture or collapse of thermostatic bellows and floats.

water (hardness) A term used to measure the extent of water supply scaling impurity commonly caused by calcium and magnesium salts. Derived carbonates and bicarbonates are generally responsible for 'temporary' hardness; chlorides, sulphates and nitrates contribute to 'permanent' hardness of water.

water scale Solid encrustation from water impurities sometimes found inside water pipes and on heat-exchanger surfaces – called 'furring' – can cause effects of water flow restriction and blockage.

water treatment A process of rendering harmless the adverse impurities in water.

water treatment – lime-soda See *Lime soda water treatment*.

water-tube boiler Developed to increase the pressure and output range of the shell boiler to about 500,000kg of steam/hour generated in one unit up to the critical steam of pressure of 217 bar.
The modern high-pressure, large-capacity water-tube boiler has been designed mainly for operation in conjunction with steam turbines for power generation, for which purpose the high pressure is essential (beneficial).
The three principal types of this boiler type are:
  1. straight-tube,
  2. bent-tube,
  3. forced-circulation.
The general construction common to the first two types comprises an assembly of water tubes connected between the steam drum(s) and the water drum(s) with the tubes expanded into the drums. The minimum number of drums is two; can be increased in number to suit the application.
The forced-circulation boiler has its heating surface principally in the form of water tubes connected to form a continuous length – (see figure 116). This boiler has extremely small tubes (32mm to 38mm diameter) and depends on a pump for the required water circulation within the boiler; it is thus independent of natural circulation which design flexibility results in a lighter and more compact boiler. Typical of this range is the La Mont forced-circulation boiler, which is used extensively in hgh-pressure, hot-water and district-heating schemes. Because of its construction, the water-tube boiler depends crucially on the highest standard of water treatment, as the scaling of a water tube will quickly result in a burst and a subsequent costly repair.
This type of boiler has wide application in power and thermal stations, in district- and group-heating schemes and in the larger industrial and oil refinery establishments where careful routine maintenance and blow down control, and water treatment is available and where the performance can be continuously monitored to anticipate potential difficulties. (See figures 116 & 117.)

194

Also see *Corner-tube boiler*.

wave energy Renewable energy obtainable from the movement of the waves across ocean surfaces. Techniques in experimental and pilot stages. Offers exciting renewable energy store once a pracicable and economic method of energy extraction has been achieved which takes due account of the very difficult environment in which the equipment has to perform effectively.

wavelength The distance between two like points on a wave shape; eg distance from crest to crest.

weather compensator control Employed to reduce fuel consumption and to provide correct conditions in space-heating systems. Compensator comprises external and pipe located detector thermostats which act through a control box to adjust the heat input to the system. May be self-acting or electrically motivated. Control may act directly on to the firing equipment (adjusting the rate of firing to the weather conditions) or actuate a mixing valve in the heat supply to the building or zone (essential when boilers provide heating and domestic hot water).

weather factor Applied to the theoretical estimation of the heat and fuel consumption of a space-heating system, based on its heating capacity, to allow for the periods during which the system operates below peak capacity due to seasonal weather variations. Varies usually in the UK between 0.6 and 0.7.

weathering Means of preventing water or moisture penetrating into a building. Lead and zinc are common roofing materials used as 'flashing' or weathering strip between a solar collector and roof tiles. Other materials used are synthetic compounds, aluminium and rubber-based strips.

weathering – equipment Method of providing a water-tight seal between mechanical or electrical components which are exposed to the external weather conditions. In many applications, such as in the locating of solar collectors, the passing of pipes, ducts and conduits

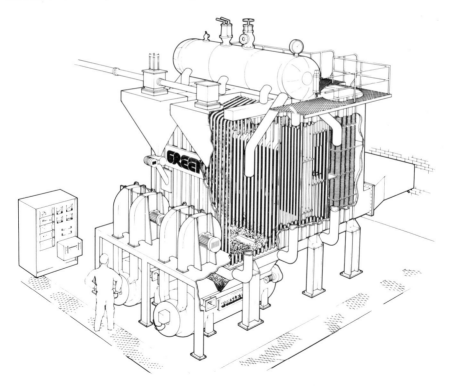

Figure 116 Green's packaged coal-fired water-tube boiler with Worsley fluidised bed combustion

through a roof construction, the door seal provided to an external air handling plant or the connection to electrical apparatus, efficient weathering to exclude all moisture may well be vital for the functioning, safety and/or life of the component or equipment.

Figures 118 to 120 show a weathering detail for the mounting on the roof of a solar collector; also the sub-frame which assists in this.

weather sensing proportional control for off-peak heating Provides the correct amount of heat as late as possible in the charge period, thus eliminating cycling losses by sensing the outside temperature. It also predicts the amount of heat required.

Application: suitable for domestic and commercial premises for new and existing off-peak systems; suitable for all tariffs.

weather-stripping Provision of seals by attachment of sealants (eg brushes, rubber flaps, etc) or caulking to and around gaps and openings (eg windows and doors) in buildings to reduce the uncontrolled infiltration of air into the building.

well – reinjection See *Reinjection well.*

wet-bulb depression The difference in readings of wet-bulb and dry-bulb thermometers for a particular sample of air.

wet-bulb temperature Temperature of air as indicated by an ordinary thermometer which has its bulb covered with a wet wick. Evaporation occurs at the wick, and the wet-bulb thermometer actually measures the temperature at which the water is evaporating. The rate at which the water evaporates at the wick relates to the prevailing relative humidity.

Wheatstone's bridge Accurate and convenient method of comparing electrical resistances by the principle of the divided circuit. Incorporated in electrical control and measurement devices.

white-meter tariff Tariff related to the time of day or night when electricity is consumed.

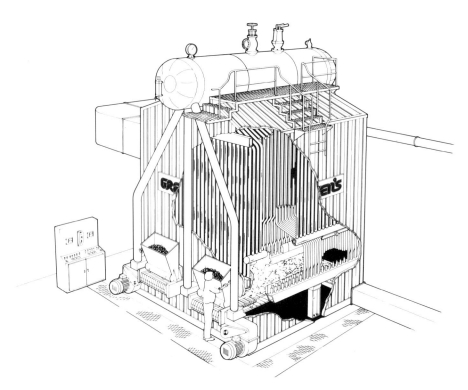

Figure 117 Green's 'A' frame water-tube packaged boiler for coal firing with chain grate stokers

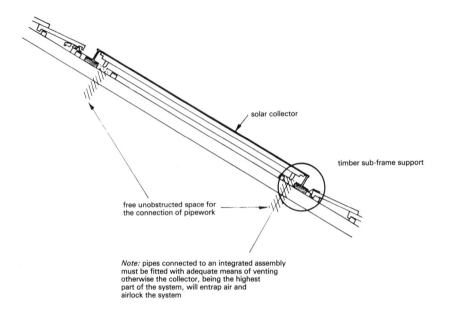

solar collector

timber sub-frame support

free unobstructed space for the connection of pipework

Note: pipes connected to an integrated assembly must be fitted with adequate means of venting otherwise the collector, being the highest part of the system, will entrap air and airlock the system

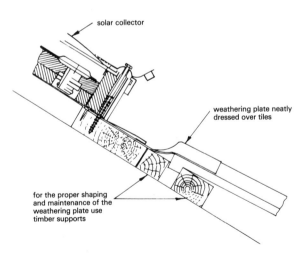

solar collector

weathering plate neatly dressed over tiles

for the proper shaping and maintenance of the weathering plate use timber supports

Figure 118 Solar collectors fitted to roof sub-frame – fixing detail

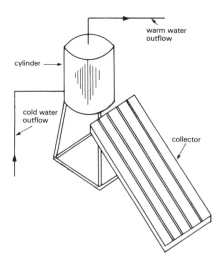

Figure 119 Single panel solar collector with hot water storage tank on flat roof location

Two-module sub-frame

Note: Carefully check weight of each assembly (including water weight) to ensure that the deflection of sub-fame (particularly with aluminium at any point does not exceed 3mm

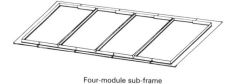

Four-module sub-frame

Figure 120 Sub-frames for solar collectors

wind farm Assembly of a number of aerogenerators on one site to feed into a common electricity off-take. Location must be carefully selected to ensure wind availability. Usually operates at wind speeds of 7 to 27 m/s (15 to 60 miles/ph).

wind generator See *Aerogenerator.*

wire-drawing See *Steam – Wire-drawing.*

working fluids Applied to heating and cooling applications. Is that fluid which carries the energy into, around or out of the system. Eg water heated in a gas-fired boiler which circulates heat to radiators and/or indirect hot water storage cylinder; refrigerant which alternately evaporates (absorbing heat) and condenses (liberating heat); air warmed by furnace, heat pump, heat exchanger, etc, circulated to a drying process or space heating; steam piped to heat exchangers, turbines etc. A system may have more than one working fluid; eg operation of heat pump, absorption refrigerator, air conditioner, etc.

Y-type valve see *Valve – Y-type.*

zoning Separation of a system (heating, ventilating, air conditioning) into different areas for purposes of improved control of the zonal environment. Zoning may be relative to solar and weather exposure, building usage, occupancy, etc.

Y
Z

Appendices

Appendix 1

Conversion Factors to SI Units

Imperial or non-SI metric unit	Multiplication factor	SI unit	Symbol
abampere	10	ampere	A
abcoulomb	10	coulomb	C(As)
abfarad	1	gigafarad	GF (As/V)
abhenry	1	nanohenry	nH (Vs/A)
abohm	1	nanoohm	$n\Omega$ (V/A)
abvolt	10	nanovolt	nV
acre	4 046.856	metre^2	m^2
angström	100	picometre	pm
atmosphere	1.013 25 $(1 \text{ bar } = 10^5 \text{ N/m}^2)$	bar	bar
barn (SI permitted unit)	10^{-28}	metre^2	m^2
barrel	0.158 987	metre^3	m^3
board foot	2.359 74	decimetre^3	dm^3
Btu (British thermal unit) IST	1.055 06	kilojoule	kJ
Btu (mean)	1.055 87	kilojoule	kJ
Btu (thermochemical)	1.054 35	kilojoule	kJ
Btu (39°F)	1.059 67	kilojoule	kJ
Btu (60°F)	1.054 68	kilojoule	kJ
bushel (UK)	36.368 7	decimetre^3	dm^3
bushel (US)	35.239 1	decimetre^3	dm^3
calorie (IST)	4.186 8	joule	J
calorie (mean)	4.190 0	joule	J
calorie (thermochemical)	4.184	joule	J
calorie (15°C)	4.185 8	joule	J
calorie (20°C)	4.181 9	joule	J
carat (new)	0.2	gramme	g
carat (old)	0.205 3	gramme	g
chain	20.116 8	metre	m
chaldron	1.309 27	metre^3	m^3
CHU	1.899 108	kilojoule	kJ
circular mil	506.707 5	picometre^2	pm^2
cord	3.624 56	metre^3	m^3
cup	236.588 2	millilitre	ml
curie (SI permitted unit)	3.7×10^{10}	disintegrations/ second	Ci
degree (angle)	17.453 29	milliradian	mrad
dyne	10	micronewton	μN
electronvolt	0.160 21	attojoule	aJ

Imperial or non-SI metric unit	Multiplication factor	SI unit	Symbol
EMU of current, resistance, etc.	same as abampere, abohm, etc.		
ESU of capacitance	1.112 6	picofarad	pF
ESU of current	333.56	picoampere	pA
ESU of potential	299.79	volt	V
ESU of inductance	0.898 76	terahenry	TH
ESU of resistance	0.898 76	teraohm	$T\Omega$
erg	100	nanojoule	nJ
Fahrenheit (degree)	$^{\circ}C = (^{\circ}F - 32) \times 5/9$ $K = (^{\circ}F - 32) \times 5/9 + 273.15$		
faraday (Carbon 12)	96.487	kilocoulomb	kC
faraday (chemical)	96.495 7	kilocoulomb	kC
faraday (physical)	96.521 9	kilocoulomb	kC
fathom	1.828 8	metre	m
fluid drachm (UK)	3.551 63	centimetre^3 (millilitre)	cm^3
fluid dram (US)	3.697	centimetre^3	cm^3
fluid minim (UK)	59.193 9	millimetre^3	mm^3
fluid minim (US)	61.609 0	millimetre^3	mm^3
fluid ounce (UK)	28.413 1	centimetre^3	cm^3
fluid ounce (US)	29.573 5	centimetre^3	cm^3
foot	0.304 8	metre	m
foot^2	0.092 903	metre^2	m^2
foot^3	0.028 316 85	metre^3	m^3
gallon (UK)	4.546 087	decimetre^3 (litre)	dm^3
gallon (US dry)	4.404 884	decimetre^3	dm^3
gallon (US liquid)	3.785 412	decimetre^3	dm^3
gamma	1	nanotesla	nT
gauss	100	microtesla	μT
gill (UK)	142.065	centimetre^3	cm^3
gill (US)	118.294	centimetre^3	cm^3
grain	0.064 798 91	gramme	g
hand	101.6	millimetre	mm
hoppus foot	0.036 054	metre^3	m^3
horsepower (boiler)	9.809 5	kilowatt	kW
horsepower (electric)	0.746	kilowatt	kW
horsepower (550 ftlb/s)	0.745 699 9	kilowatt	kW
horsepower (metric)	0.735 499	kilowatt	kW
horsepower (UK)	0.745 7	kilowatt	kW
horsepower (water)	0.746 43	kilowatt	kW
hundredweight (UK)	50.802 35	kilogramme	kg
hundredweight (US)	45.359 24	kilogramme	kg

Imperial or non-SI metric unit	Multiplication factor	SI unit	Symbol
inch	25.4 (exactly)	millimetre	mm
inch^2	645.16	millimetre^2	mm^2
inch^3	16.387 06	centimetre^3	cm^3
iron	0.53	millimetre	mm
kilogramme force	9.806 65	newton	N
kip	4.448 222	kilonewton	kN
knot	0.514 444 4	metre/second	m/s
lambert	3.183 099	kilocandela/metre^2	kcd/m^2
langley	41.84	kilojoule/metre^2	kJ/m^2
league (international nautical)	5 556.000	metre	m
league (statute)	4 828.032	metre	m
league (UK nautical)	5 559.552	metre	m
lightyear (SI permitted unit)	9 460.55	terametre	Tm
link	0.201 168	metre	m
maxwell	10	nanoweber	nWb
micron	1	micrometre	μm
mil	25.4	micrometre	μm
mile (nautical UK)	1 853.184	metre	m
mile (nautical US)	1 852.000	metre	m
mile (statute)	1 609.344	metre	m
mile^2	2.589 988	kilometre^2	km^2
minute (angle)	290.888 2	microradian	μrad
oersted	79.577 47	ampere/metre	A/m
ounce force	0.278 013 9	newton	N
ounce mass (avoirdupois)	28.349 52	gramme	g
ounce troy	31.103 48	gramme	g
parsec (SI permitted) unit)	$3.083\ 74 \times 10^{16}$	metre	m
peck (UK)	9.092 18	decimetre^3	dm^3
peck (US)	8.809 768	decimetre^3	dm^3
pennyweight	1.555 174	gramme	g
perch (masonry)	0.701	metre^3	m^3
Petrograd standard (timber)	4.672 28	metre^3	m^3
phot	10	kilolumen/metre^2	klm/m^2
pica (printer's)	4.217 518	millimetre	mm
pint (UK)	0.568 261	decimetre^3	dm^3
pint (US dry)	0.550 610 5	decimetre^3	dm^3
pint (US liquid)	0.473 176 5	decimetre^3	dm^3
point (printer's)	0.351 459	millimetre	mm
point (silversmith's)	6.4	micrometre	μ

Imperial or non-SI metric unit	Multiplication factor	SI unit	Symbol
poise (SI permitted unit)	0.1	newton second/ metre^2	N s/m^2
poundal	0.138 255	newton	N
pound (force)	4.448 222	newton	N
pound (mass)	0.453 592 4	kilogramme	kg
pound (troy)	0.373 241 7	kilogramme	kg
quart (UK)	1.136 52	decimetre^3	dm^3
quart (US dry)	1.101 221	decimetre^3	dm^3
quart (US liquid)	0.946 353	decimetre^3	dm^3
quarter (UK)	12.700 6	kilogramme	kg
Rankine degree	1.8	degree Kelvin	K
rod	5.029 2	metre	m
roentgen	0.257 976	millicoulomb/ kilogramme	mC/kg
second (angle)	4.848 137	microradian	μrad
slug	14.593 9	kilogramme	kg
Statampere, statohm, etc., same as ESU of current, resistance, etc.			
stokes (SI permitted unit)	1	centimetre^2/ second	cm^2/s
stone (UK)	6.350 29	kilogramme	kg
tablespoon	14.786 76	centimetre^3	cm^3
teaspoon	4.928 922	centimetre^3	cm^3
ton (assay)	29.166 67	gramme	g
ton (UK long)	1 016.047	kilogramme	kg
ton (US short)	907.184 7	kilogramme	kg
ton (nuclear)	4.20	gigajoule	GJ
ton (refrigeration)	3.516 85	kilowatt	kW
ton (register)	2.831 685	metre^3	m^3
ton (shipping UK)	1.189 3	metre^3	m^3
ton (shipping US)	1.132 67	metre^3	m^3
ton (displacement)	0.991 1	metre^3	m^3
torr	133.322	newton/metre^2	N/m^2
township (US)	93.239 57	kilometre^2	km^2
unit pole	125.663 7	nanoweber	nWb
yard	0.914 4	metre	m
yard^2	0.836 127 4	metre^2	m^2
yard^3	0.764 554 9	metre^3	m^3

Appendix 2

Combined units

It is impossible to give all the combinations of units which can be built up from the values in Appendix 1. The required combined units can, however, easily be calculated from the simple units.

Many of the more common compound conversions are given in tables 1 to 5.

Table 1: mass per unit length, area and volume

Imperial or non-SI metric unit	Multiplication factor	SI unit	Symbol
ton (UK)/mile	0.631 342	kilogramme/ metre	kg/m
ton (US)/mile	0.563 698	kilogramme/ metre	kg/m
pound/yard	0.496 055	kilogramme/metre	kg/m
pound/foot	1.488 16	kilogramme/metre	kg/m
pound/inch	17.858	kilogramme/metre	kg/m
ton (UK)/mile^2	392.298	kilogramme/ kilometre^2	kg/km^2
ton (US)/mile^2	350.266	kilogramme/ kilometre^2	kg/km^2
pound/foot^2	4.882 43	kilogramme/metre^2	kg/m^2
ounce/foot^2	0.305 152	kilogramme/metre^2	kg/m^2
ton (UK)/yard^3	1 328.94	kilogramme/metre^3	kg/m^3
ton (US)/yard^3	1 186.55	kilogramme/metre^3	kg/m^3
pound/foot^3	16.018 5	kilogramme/metre^3	kg/m^3
pound/gallon (UK)	0.099 776	kilogramme/ decimetre^3	kg/dm^3
pound/gallon (US)	0.119 826 4	kilogramme/ decimetre^3	kg/dm^3
pound/inch^3	27 679.9	kilogramme/metre^3	kg/m^3

Table 2: force per unit length, moments and momentum

Imperial or non-SI metric unit	Multiplication factor	SI unit	Symbol
ton (UK) force/foot	32.690 3	kilonewton/metre	kN/m
ton (US) force/foot	29.187 8	kilonewton/metre	kN/m
pound force/foot	14.593 9	newton/metre	N/m
pound force/inch	175.127	newton/metre	N/m
ton (UK) force foot	3.037 03	kilonewton metre	kNm
ton (US) force foot	2.711 63	kilonewton metre	kNm
pound force foot	1.355 82	newton metre	Nm

Imperial or non-SI metric unit	Multiplication factor	SI unit	Symbol
pound force inch	0.112 985	newton metre	Nm
pound foot^2	0.042 140	kilogramme metre^2	kgm^2
pound inch^2	2.926 4	kilogramme centimetre^2	kg cm^2
pound foot/second	0.138 255	kilogramme metre/second	kg m/s

Table 3: pressures

Imperial or non-SI metric unit	Multiplication factor	SI unit	Symbol
centimetre of mercury (0°C)	1.333 22	kilonewton/metre^2	kN/m^2
centimetre of water (4°C)	98.063 8	newton/metre^2	N/m^2
foot of water (39°F)	2.988 98	kilonewton/metre^2	kN/m^2
inch of mercury (32°F)	3.386 389	kilonewton/metre^2	kN/m^2
inch of water (39°F)	249.082	newton/metre^2	N/m^2
kilogramme force/metre^2	9.806 65	newton/metre^2	N/m^2
poundal/foot^2	1.488 164	newton/metre^2	N/m^2
pound force/foot^2	47.880 26	newton/metre^2	N/m^2
pound force/inch^2	6.894 757	kilonewton/metre^2	kN/m^2
ton (UK) force/foot^2	107.252	kilonewton/metre^2	kN/m^2
ton (US) force/foot^2	95.760 7	kilonewton/metre^2	kN/m^2
ton (UK) force/inch^2	15.444 3	newton/millimetre^2	N/mm^2 or MN/m^2
ton (US) force/inch^2	13.789 55	newton/millimetre^2	N/mm^2 or MN/m^2

Table 4: heat and heat transfer

Imperial or non-SI metric unit	Multiplication factor	SI unit	Symbol
Btu/hour	0.293 071	watt	W (J/s)
Btu/pound	2.326	kilojoule/ kilogramme	kJ/kg
Btu/foot^3	37.258 9	kilojoule/metre^3	kJ/m^3
therm/UK gallon	23.208 0	gigajoule/metre^3	GJ/m^3
therm/US gallon	27.871 7	gigajoule/metre^3	GJ/m^3
Btu/pound °F	4.186 8	kilojoule/ kilogramme K	kJ/kg K

Imperial or non-SI metric unit	Multiplication factor	SI unit	Symbol
Btu/foot³ °F	67.066 1	kilojoule/metre³ K	kJ/m³ K
Btu/foot² hour	3.154 59	watt/metre²	W/m²
Btu/foot² hour °F	5.678 26	watt/metre² deg C	W/m² deg C
kilocalorie/metre² hour °C	1.163	watt/metre² deg C	W/m² deg C
Btu/foot hour °F	1.730 073	watt/metre deg C	W/m deg C
Btu inch/foot² hour °F	0.144 228	watt/metre deg C	W/m deg C
Btu inch/foot² second °F	519.22	watt/metre deg C	W/m deg C
kilocalorie/metre hour °C	1.163	watt/metre deg C	W/m deg C
calorie/centimetre second °C	418.68	watt/metre deg C	W/m deg C

Table 5: liquids

Imperial or non-SI metric unit	Multiplication factor	SI unit	Symbol
grain/100 foot³	0.022 883 5	gramme/metre³	g/m³
ounce/UK gallon	6.236 02	gramme/decimetre³	g/dm³
ounce/US gallon	7.489 142	gramme/decimetre³	g/dm³
pound force hour/foot²	0.172 369	meganewton second/metre²	MNs/m²
pound force second/foot²	47.880 3	newton second/metre²	Ns/m²
poundal second/foot²	1.488 16	newton second/metre²	Ns/m²
foot²/hour	25.806 4	millimetre²/second	mm²/s
foot²/second	0.092 903	metre²/second	m²/s
poise	0.1	newton second/metre²	Ns/m²
stoke	1	centimetre²/second	cm²/s

Appendix 3

Section 1: British standard symbols for pipelines & fittings BS 2917, BS 974, BS M24

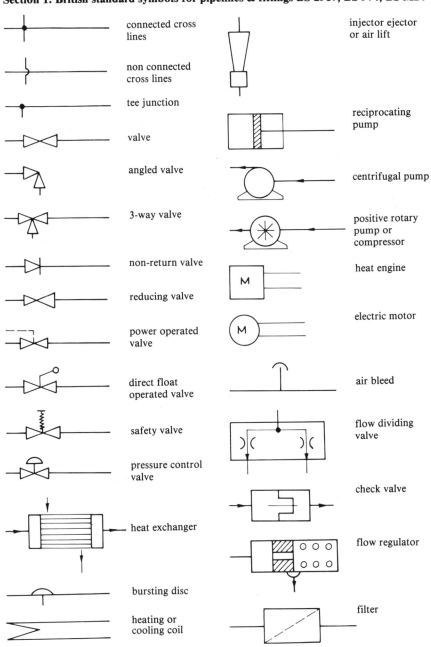

connected cross lines	injector ejector or air lift
non connected cross lines	
tee junction	reciprocating pump
valve	
angled valve	centrifugal pump
3-way valve	positive rotary pump or compressor
non-return valve	heat engine
reducing valve	
power operated valve	electric motor
direct float operated valve	air bleed
safety valve	flow dividing valve
pressure control valve	
heat exchanger	check valve
	flow regulator
bursting disc	
heating or cooling coil	filter

Appendix 3

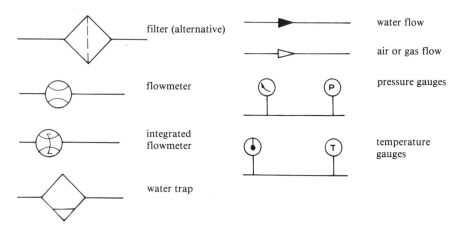

filter (alternative)	water flow
flowmeter	air or gas flow
integrated flowmeter	pressure gauges
water trap	temperature gauges

Section 2: European standard symbols for pipelines & fittings
DIN-2481, DIN-1988, DIN-2429

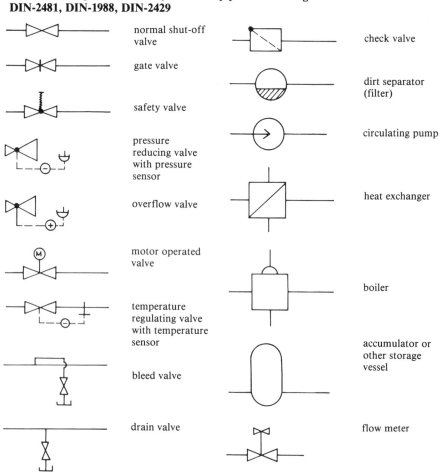

normal shut-off valve	check valve
gate valve	dirt separator (filter)
safety valve	circulating pump
pressure reducing valve with pressure sensor	heat exchanger
overflow valve	boiler
motor operated valve	accumulator or other storage vessel
temperature regulating valve with temperature sensor	flow meter
bleed valve	
drain valve	

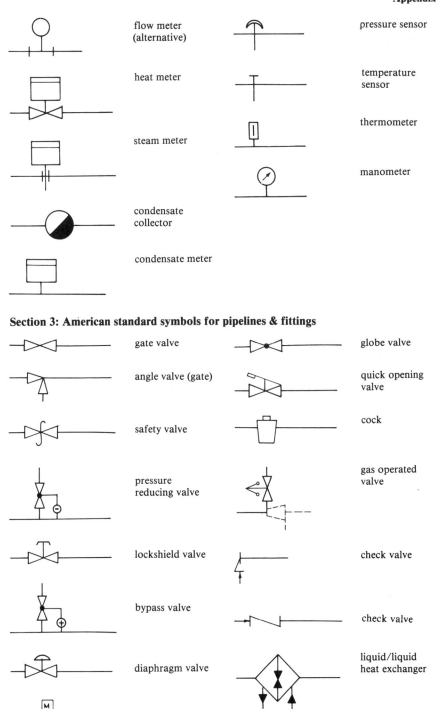

flow meter (alternative)

pressure sensor

heat meter

temperature sensor

steam meter

thermometer

manometer

condensate collector

condensate meter

Section 3: American standard symbols for pipelines & fittings

gate valve

globe valve

angle valve (gate)

quick opening valve

safety valve

cock

pressure reducing valve

gas operated valve

lockshield valve

check valve

bypass valve

check valve

diaphragm valve

liquid/liquid heat exchanger

motorized valve

213

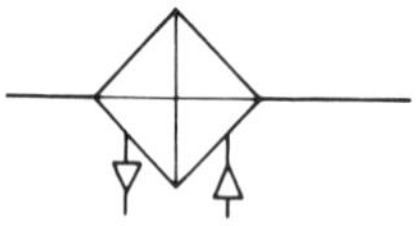

gas/liquid heat exchanger

compressor (alternative)

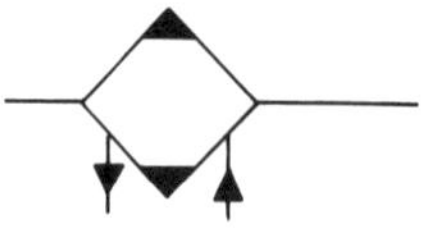

liquid/liquid/ cooler

heat engine

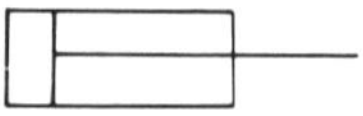

compressor

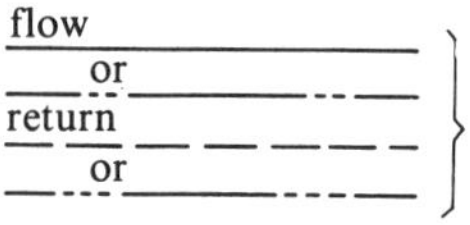

flow
or
return
or

} hot water

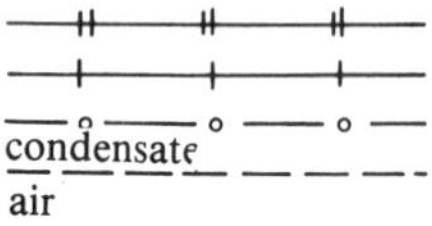

steam { high pressure
        low pressure

condensate
air

NOTE: If other fluids are being pumped an uninterrupted line is used for flow. A dashed line for return with code letters in centre as under

———————— CHF ————————

chilled water flow

———————— CHR ————————

chilled water return

Other abbreviations used are:

B – brine flow
CR – condenser return
A – compressed air
FOF – fuel oil flow
G – gas
BR – brine return
D – drain

RL – refrigeration liquid
FOR – fuel oil return
V – vacuum
C – condenser water
H – humidification line
RD – refrigeration discharge
FOV – fuel oil vent